Scaling the Wall

A Survey of the Riemann Wall,
and the Walls Beside It

Book Three of The Primes Trilogy

John Maggio
Efficient Intelligence, LLC

Every method begins at the wheel, the primes' local law,
and climbs to the wall, their global secret.
This book is the climb, and the honest report from the top.

Published by Efficient Intelligence, LLC

Paperback ISBN: 979-8-9967325-6-2
Hardcover ISBN: 979-8-9967325-7-9
Ebook ISBN: 979-8-9967325-8-6

First edition, 2026 Printed in the United States of America

Advance in Peace

Preface to the Trilogy

Three books stand behind this one's argument, and this is the third. The first, *A Wheel Through the Primes*, built the wheel: the local law of the primes, the structure of the residues modulo $210 = 2 \cdot 3 \cdot 5 \cdot 7$, the forty-eight survivors coprime to it, the subprimes that remain, and the Euler product that turns all of this into a function. The second, *A Spectrum Through the Primes*, followed that local law upward into the zeros—the spectrum hidden inside the prime counting function—and the family of L-functions the wheel generates. This third book is the advanced, self-contained survey. It takes a dozen of the great approaches to the Riemann Hypothesis, grows each one honestly from the wheel, climbs it to the wall, and delivers, in the same chapter, the exact place and reason each approach is turned back.

The discipline of the trilogy is honesty, and it is strict. Real mathematics is credited by the names of those who made it. Metaphor is marked as metaphor and never allowed to do the work of a theorem. What is proven is kept rigorously apart from what is open, and the structure that lures a method forward is never mistaken for a lever that crosses the wall. Every chapter carries a computation—standard-library Python, deterministic, printed in full—so that its claims can be checked rather than trusted.

This book is self-contained. Where it needs the wheel, the Euler product, the explicit formula, or the L-functions, it re-teaches them in its own language and its own figures. A reader who has never opened the first two books can read this one from the first page; a reader who has will recognise old friends in new dress.

The author gratefully acknowledges the use of Anthropic's Claude as a computational and drafting assistant during the preparation of this work. Any mathematical results or interpretations presented here remain entirely the responsibility of the author.

Introduction: Scaling the Wall

The thesis of this book can be put in one sentence: *the primes surrender their local structure freely and their global structure not at all*, and the Riemann Hypothesis is the sharpest statement of that refusal. The wheel—residues, densities, the singular series, the characters—is the local law, and it is known. Where the zeros lie, whether two primes recur forever, how to undo a product: these are the global facts, and they are walled. Every chapter here approaches the wall along a different road, and every road runs out at the same kind of edge.

Each chapter has one shape. It begins with a *lure*: a genuine method, grown from the wheel, that seems for a while to lead over the wall. It *climbs*: the method is developed with its real equations and its honest gains. And it *turns*: in the same chapter, the deflation is delivered—the precise point at which the method is beaten back, and the reason. The reader is lured in, but never deceived.

The book is in three parts. **Part I, The Wheel and the Wall**, lays the foundation—the wheel, the Euler product, the mirror, the explicit formula, the wall itself—and then climbs one floor up to Ramanujan's τ, where the analogue of the Hypothesis is a proven theorem. **Part II, The Five Faces of the Hypothesis**, takes five reformulations, each *provably equivalent* to the Riemann Hypothesis—the oscillation of the prime-counting error, the positivity of Li's and Weil's functionals, the completeness of a span of fractional parts, an extremal inequality on the divisor sum, and the constant of a heat flow—and shows each one to be the same wall seen from a new side. **Part III, The Walls**, steps back: it visits the neighbouring walls that ring the Hypothesis (the random-matrix statistics of the zeros, the undiscovered operator, the weaker Lindelöf shadow, and the geometry where the Hypothesis is already a theorem), walks the twin-prime corridor to its parity barrier, and closes with the three great walls together—location, recurrence, and decomposition—and the single situation that runs through all three.

The closing chapter states the unifying lesson plainly: at each wall, the lever that would cross it turns out to be the very thing being sought. To control the zeros one needs the zeros; to isolate two primes one needs to break parity, which is to isolate two primes; to factor cheaply one needs to undo multiplication, which is to factor. The destination is the tool. That circularity is not a failure of ingenuity—it is the structure of the subject, and to map it exactly is the work of this book.

A word on the computations. Each chapter ends with a verification: a short, deterministic program using only the Python standard library, which checks the chapter's claims to a stated range and prints its results. They are reproductions, not decorations—the reader can run them. Where a fact is elementary it is checked exactly; where it is deep, a faithful model or a finite instance is checked, and the limits of the check are stated. Plain language and full equations stand side by side throughout, and the figures are meant to carry the argument as much as the prose.

The wheel turns at the foot of the wall. What follows is the climb.

Contents

Part I

The Wheel and the Wall

Chapter 1

The Wheel and the Wall

This is the foundation of an advanced book, and its one job is to make the rest self-contained. It assembles, in one place and in this book's own language and pictures, everything the later chapters stand on: the wheel as the Euler product—the base*— the family of L-functions the wheel carries, the mirror that is the functional equation, the explicit formula that is the hinge between the primes and the zeros, and the Riemann wall that every later chapter will climb toward and be turned back from. None of it is new. All of it is classical, and the two books before this one develop it slowly and with care. The reason to gather it here, at speed and in a single vantage, is that what follows is a survey—a broad sweep over how the Riemann Hypothesis has been attacked, and how each attack was beaten back— and a survey needs a place to stand. This chapter is that place. A reader who has the earlier books can move through it quickly; a reader who does not can still follow everything after it.*

Scope

This chapter consolidates the two earlier volumes. It assumes only undergraduate analysis and a first course in number theory; everything beyond that it builds in passing. Each section states an object the rest of the book will use—the wheel, the Euler product, the family, the functional equation, the explicit formula, the wall— gives the equation, says in plain words what the equation means, and points to where the earlier books treat it in full. The value here is not novelty but vantage: one base, one language, and one picture (Figure 1.2) that the entire survey hangs from.

1.1 The wheel

Start with the primes' simplest secret: most numbers cannot be prime, and you can see which ones at a glance. A number divisible by 2, 3, 5, or 7 is composite (past those primes themselves), so every prime beyond 7 must fall in a residue class modulo $210 = 2 \cdot 3 \cdot 5 \cdot 7$ that shares no factor with 210. There are exactly

$$\varphi(210) = 210\Big(1 - \tfrac{1}{2}\Big)\Big(1 - \tfrac{1}{3}\Big)\Big(1 - \tfrac{1}{5}\Big)\Big(1 - \tfrac{1}{7}\Big) = 48 \tag{1.1}$$

such classes. Lay the numbers 1 to 210 around a circle and keep only those 48 spokes: that is *the wheel.* Every prime past 7 lands on a spoke; the other 162 positions are barred to primes forever. The wheel is the primes' local law—a statement about residues, decided by the small primes alone, that holds for all integers at once.

The surviving spokes are not all prime. The number 1 sits on one. The primes from 11 up sit on others. And some spokes carry composites that simply have no small factor—numbers like $121 = 11^2$, $143 = 11 \cdot 13$, $169 = 13^2$. Following the earlier books we call these the *subprimes*: composite, coprime to 210, least prime factor at least 11. The first is 121. The wheel cannot tell a prime from a subprime—both are survivors—and that blindness, harmless here, will reappear at one of the walls.

Nothing about 210 is special except convenience. Multiply in the next prime and the wheel grows: $2 \to 6 \to 30 \to 210 \to 2310 \to \cdots$, the primorials, each a finer sieve than the last. The wheel is a family of sieves, and which one you use is a matter of how much resolution you want.

1.2 The Euler product: the wheel becomes a function

Now turn the wheel into a single analytic object. For a complex number s with $\operatorname{Re} s > 1$ define the Riemann zeta function by the sum over all integers,

$$\zeta(s) = \sum_{n=1}^{\infty} \frac{1}{n^s}. \tag{1.2}$$

Because every integer factors uniquely into primes, this sum equals a product over primes—Euler's identity:

$$\zeta(s) = \prod_{p} \frac{1}{1 - p^{-s}}. \tag{1.3}$$

The reason is one line. Expand each factor as a geometric series, $\left(1 - p^{-s}\right)^{-1} = 1 + p^{-s} + p^{-2s} + \cdots$, multiply the factors over all primes, and every term of the product is $\left(p_1^{a_1} \cdots p_k^{a_k}\right)^{-s} = n^{-s}$ for exactly one integer n, by unique factorization. The sum over integers and the product over primes are the same object.

Equation (1.3) is the wheel made into a function. Each prime contributes one factor; the wheel modulo 210 is what you get by stripping off the first four of them, the factors for $2, 3, 5, 7$. In plain terms: ζ is the primes, multiplied together. Everything in this book that looks analytic—zeros, critical lines, explicit formulas—enters only through this product, which is why the wheel sits at the base of all of it. The verification confirms the identity numerically, $\prod_p (1 - p^{-2})^{-1} = \zeta(2) = \pi^2/6$, and recovers the wheel's survivor density $\frac{48}{210} = \prod_{p|210}(1 - 1/p)$ as the product of its local factors.

1.3 The family the wheel carries

The wheel does not carry ζ alone. To each residue class one can attach a *Dirichlet character* χ, a function on the integers that is periodic modulo the wheel and multiplicative, $\chi(mn) = \chi(m)\chi(n)$. Each character has its own zeta-like function, again with an Euler product,

$$L(s,\chi) = \sum_{n=1}^{\infty} \frac{\chi(n)}{n^s} = \prod_p \frac{1}{1-\chi(p)\,p^{-s}}. \tag{1.4}$$

The ordinary zeta function is the case of the trivial character, $\chi \equiv 1$. The wheel of 210 carries a whole family of these—among them seven nontrivial real characters, indexed by the seven nontrivial divisors of $105 = 3 \cdot 5 \cdot 7$, which the first book draws as a Boolean lattice of conductors. We will not need the lattice in detail here; what matters for the survey is that the wheel's finer structure—the race between residue classes, the cancellations in its sums—is governed not by ζ alone but by this entire family. That is why several attacks in this book run not into the Riemann Hypothesis but into its generalization, the same statement asked of every $L(s,\chi)$ at once.

1.4 The mirror: the functional equation

The zeta function, defined by the sum only for $\operatorname{Re} s > 1$, extends to the whole complex plane, and the extension has a symmetry. Completing ζ with the factors that belong to the prime "at infinity,"

$$\xi(s) = \tfrac{1}{2}\, s(s-1)\, \pi^{-s/2}\, \Gamma\!\left(\tfrac{s}{2}\right) \zeta(s), \tag{1.5}$$

produces an entire function obeying the clean reflection

$$\xi(s) = \xi(1-s). \tag{1.6}$$

In plain words: the completed function is unchanged when you reflect the plane across the vertical line $\operatorname{Re} s = \frac{1}{2}$. That line is the *critical line*, the mirror's axis. The zeta function has "trivial" zeros at the negative even integers $-2, -4, -6, \ldots$, forced by the poles of the gamma factor, and all its other—"nontrivial"—zeros ρ lie in the critical strip $0 < \operatorname{Re} s < 1$. By the symmetry (1.6) they come in pairs $\rho, 1-\rho$, placed mirror-image across the critical line. The first book reaches this same reflection from the other side, showing that the wheel's geometric mirror $r \mapsto 210 - r$ and the analytic mirror $s \mapsto 1 - s$ are one fact, hinged by the parity $\chi(-1)$.

1.5 The hinge: the explicit formula

Here is the object the whole book swings on. Count the primes in a weighted way with the Chebyshev function

$$\psi(x) = \sum_{p^k \le x} \log p, \tag{1.7}$$

a staircase that jumps by $\log p$ at every prime power. Riemann and von Mangoldt connected this staircase to the zeros by an exact identity, the *explicit formula*:

$$\psi(x) = x - \sum_{\rho} \frac{x^{\rho}}{\rho} - \log(2\pi) - \tfrac{1}{2}\log(1 - x^{-2}), \tag{1.8}$$

the sum running over all nontrivial zeros ρ. Read it slowly, because everything later depends on it. The left side is built from primes; the right side is built from zeros. The leading term x is the average growth; every wobble of the staircase around that average is the sum $\sum_\rho x^\rho/\rho$ over the zeros. Pairing a zero $\rho = \frac{1}{2} + i\gamma$ with its mirror $\frac{1}{2} - i\gamma$, each pair contributes a wave $2\sqrt{x}\,\big(\frac{1}{2}\cos(\gamma\log x) + \gamma\sin(\gamma\log x)\big)/(\frac{1}{4}+\gamma^2)$, an oscillation whose size is set by $\sqrt{x}$ when the zero sits on the critical line.

So the primes and the zeros are not two subjects but two readings of one identity. The zeros are the frequencies; the primes are the tune. Figure 1.1 shows this literally: feed the first sixty zeros into Eq. (1.8) and the staircase reassembles, jump by jump, sharpening as more zeros are added. The verification measures the agreement—a correlation above 0.9 from sixty zeros. This is the hinge. Every attack in the chapters to come is, at bottom, an attempt to control the left side by controlling the right, or the reverse, and the explicit formula is the door each one walks through.

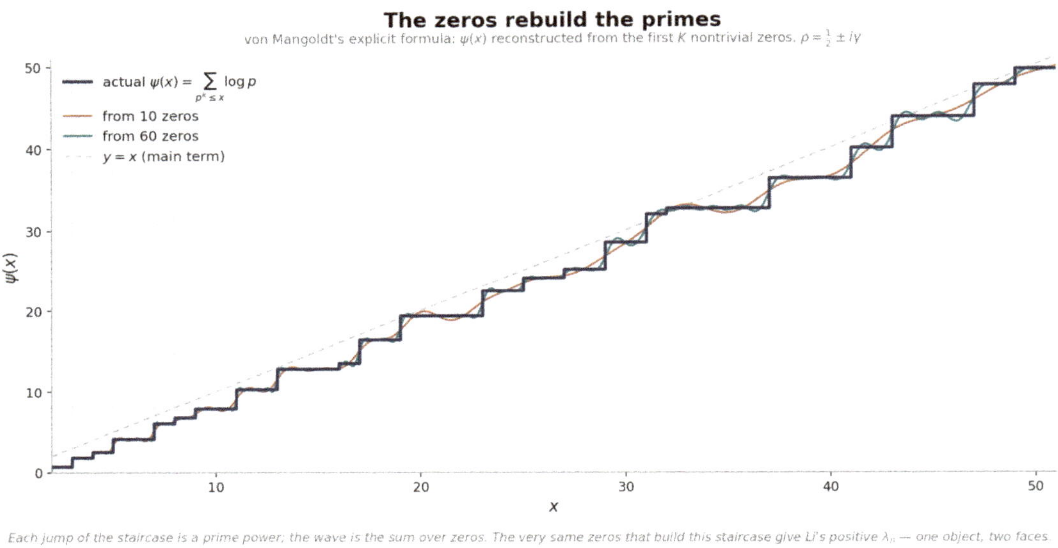

Figure 1.1: The hinge in one picture. Von Mangoldt's explicit formula (1.8) reconstructs the prime staircase $\psi(x)$ from the first K nontrivial zeros $\rho = \frac{1}{2} \pm i\gamma$: ten zeros give the slow swell, sixty trace the steps. The primes are downstream of the zeros, and this is what that means.

1.6 The wall

Where do the zeros sit? In the strip, in conjugate pairs, mirror-symmetric about the critical line. The *Riemann Hypothesis* is the claim that they sit *on* it:

$$\text{every nontrivial zero } \rho \text{ has } \operatorname{Re}\rho = \tfrac{1}{2}. \tag{1.9}$$

By the explicit formula this is not an idle statement about a function; it is the assertion that the primes are as regular as arithmetic allows. If every ρ has real part $\frac{1}{2}$, then

$|x^\rho| = \sqrt{x}$ and the wobble in Eq. (1.8) is held to the smallest possible size, giving $\psi(x) = x + O(\sqrt{x}\,(\log x)^2)$; if even one zero drifts off the line, the primes fluctuate more wildly than that. The Hypothesis is the wall this book is named for, and Figure 1.2 is the book's map: the wheel and its Euler product at the foot, the family and the mirror and the hinge built on top, and the wall at the summit.

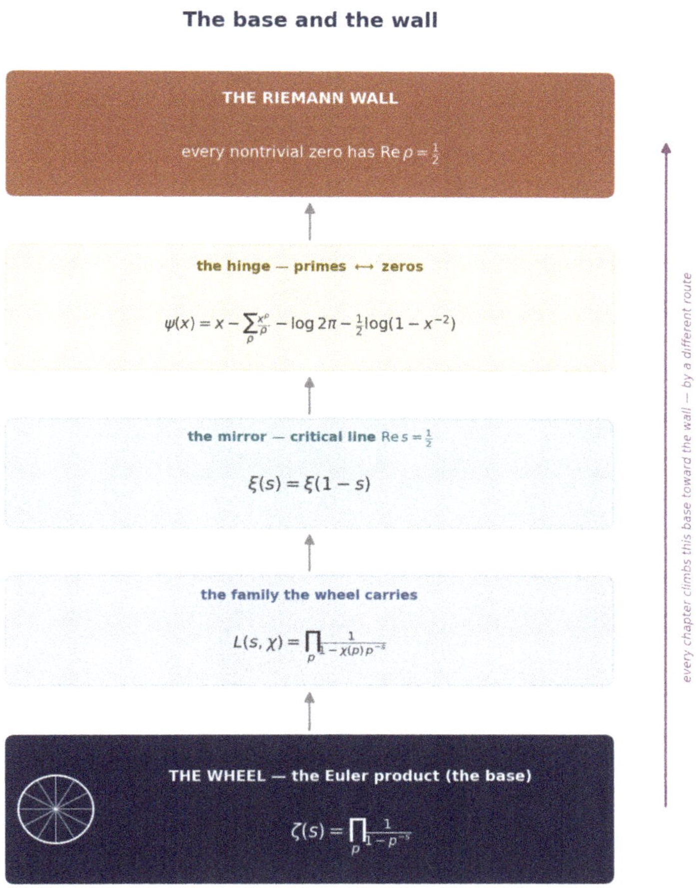

Figure 1.2: The base and the wall. The wheel—the Euler product—is the foundation; the family of L-functions, the functional-equation mirror, and the explicit-formula hinge are built on it; the Riemann Hypothesis is the wall at the top. Every chapter that follows climbs this same base toward the same wall, by a different route.

The Hypothesis is one wall of three the books circle. The second is the twin primes: whether two primes two apart recur forever, a question whose density the wheel hands over and whose existence it does not. The third is factoring: that a number is easy to build from primes and hard to break back into them. All three rise from the same base. This volume concentrates on the first and its many disguises, with the other two standing at the close.

1.7 The plan: climbing the base

The rest of the book is a single motion, repeated. Each chapter takes one genuine attack on the Riemann wall—a real method, named and credited—starts it from the base just assembled, and follows it up until it stops. We will watch the lure of each method completely, granting it every real thing it offers, and then name, at the top, exactly where and why it is turned back. The turns, chapter to chapter, will start to rhyme.

The climb begins one floor up from ζ, with Ramanujan's τ and its L-function, to see what changes when the object over the wheel has rank two instead of one. Then the survey proper: the reformulations of the Hypothesis, taken by *face*. The oscillation face—the prime number theorem's error term, the Mertens and Liouville sums—where the cancellation one must prove *is* the zeros, and where two beautiful strengthenings turned out to be false. The positivity face, Li's criterion and Weil's, where the wall becomes a sequence that must stay non-negative. The completeness face, Nyman and Beurling's, where it becomes an approximation in a Hilbert space. The extremal face, Robin's and Lagarias's inequalities on the divisor function. The regularity face, the Laguerre–Pólya class and the de Bruijn–Newman constant. Then the neighbouring walls—the spacing of the zeros, the size of ζ on the line, the search for an operator, the geometry that proves the Hypothesis in another world—each reached from the same wheel and each a different barrier. And at the end, the three walls together.

Every one of these will be written the same way: the full equations, the plain-language reading of what they say, and the honest account of the wall they meet—which, the survey will show, is in the end one wall wearing many faces.

What this is and is not

What this chapter establishes is genuine and entirely classical: the Euler product, the Dirichlet family, the functional equation, the explicit formula, and the statement of the Hypothesis, assembled into one base with one set of pictures so that the rest of the book can stand alone. The verification checks the load-bearing pieces—that the Euler product is ζ, that the wheel is its first factors, that the explicit formula rebuilds the primes from the zeros.

What it is not is a step toward crossing the wall. Nothing here is new, and nothing here is meant to be; this is the ground, deliberately. Its only claim is that everything the survey attacks rises from exactly this foundation, so that when each later method is turned back, we will be able to see the turn against a base we already hold in common.

Verification

The script below, standard library only and deterministic, checks the foundation: the Euler product equals ζ at $s = 2$ and $s = 4$; the wheel's survivor density is the product of its local Euler factors, $\varphi(210)/210 = 48/210$; the 48 survivors are the unit, the primes from 11 up, and the subprimes (the first being $121 = 11^2$); and the explicit formula rebuilds the prime staircase from the first sixty zeros.

```
#!/usr/bin/env python3
"""Verification for 'The Wheel and the Wall' (Book Three, Chapter 1).
Standard library only; deterministic. Establishes the base: the Euler product IS
the wheel made into a function, and the explicit formula is the hinge to the zeros."""
import math

checks=0
def ok(cond,msg):
    global checks
    assert cond,"FAIL: "+msg
    checks+=1; print("PASS:",msg)

def primes_upto(n):
    s=bytearray([1])*(n+1); s[0]=s[1]=0; i=2
    while i*i<=n:
        if s[i]: s[i*i::i]=bytearray(len(s[i*i::i]))
        i+=1
    return [i for i in range(2,n+1) if s[i]]

# ===== the Euler product IS zeta: prod_p (1 - p^-s)^-1 = sum_n n^-s =====
P=primes_upto(200000)
def euler(s):
    pr=1.0
    for p in P: pr*=1.0/(1.0-p**(-s))
    return pr
z2=euler(2); z4=euler(4)
print(f"  Euler product s=2 -> {z2:.6f}  (pi^2/6={math.pi**2/6:.6f})")
print(f"  Euler product s=4 -> {z4:.6f}  (pi^4/90={math.pi**4/90:.6f})")
ok(abs(z2-math.pi**2/6)<1e-4, "Euler product over primes equals zeta(2)=pi^2/6")
ok(abs(z4-math.pi**4/90)<1e-6, "Euler product over primes equals zeta(4)=pi^4/90")

# ===== the wheel is the first Euler factors: density = prod_{p|210}(1-1/p) =====
wheel_primes=[2,3,5,7]
dens=1.0
for p in wheel_primes: dens*=(1-1.0/p)
phi210=sum(1 for r in range(1,211) if math.gcd(r,210)==1)
print(f"  wheel survivor density = {dens:.6f} = {phi210}/210; phi(210)={phi210}")
ok(abs(dens-phi210/210.0)<1e-12 and phi210==48,
   "wheel density prod(1-1/p)=phi(210)/210=48/210 (local Euler factors)")

# the survivors split into the unit, the primes >=11, and the subprimes (lpf>=11)
surv=[r for r in range(1,211) if math.gcd(r,210)==1]
primeset=set(primes_upto(210))
prime_surv=[r for r in surv if r in primeset]
subprime=[r for r in surv if r not in primeset and r!=1]
ok(surv[0]==1 and min(subprime)==121,
   "the 48 survivors: unit 1, primes >=11, and subprimes (first is 121=11^2)")

# ===== the hinge: the explicit formula rebuilds the primes from the zeros =====
GAMMA=[14.13472514,21.02203964,25.01085758,30.42487613,32.93506159,37.58617816,
 40.91871901,43.32707328,48.00515088,49.77383248,52.97032148,56.4462477,59.347044,
 60.83177852,65.11254405,67.07981053,69.54640171,72.06715767,75.7046907,77.14484007,
 79.33737502,82.91038085,84.73549298,87.42527461,88.80911121,92.49189927,94.65134404,
 95.87063423,98.83119422,101.31785101,103.72553804,105.44662305,107.16861118,
 111.02953554,111.87465918,114.32022092,116.22668032,118.79078287,121.370125,
 122.94682929,124.25681855,127.51668388,129.5787042,131.08768853,133.4977372,
 134.75650975,138.11604205,139.73620895,141.1237074,143.11184581,146.00098249,
 147.42276534,150.05352042,150.92525761,153.02469381,156.11290929,157.59759182,
 158.84998817,161.18896414,163.03070969]
def psi_zeros(x):
```

```
    L=math.log(x); s=0.0
    for g in GAMMA: s+=(0.5*math.cos(g*L)+g*math.sin(g*L))/(0.25+g*g)
    return x-math.log(2*math.pi)-0.5*math.log(1-x**-2)-2*math.sqrt(x)*s
def psi_actual(x):
    t=0.0
    for p in primes_upto(int(x)):
        q=p
        while q<=x: t+=math.log(p); q*=p
    return t
xs=[2+0.2*k for k in range(1,490)]
A=[psi_actual(x)-x for x in xs]; B=[psi_zeros(x)-x for x in xs]
ma=sum(A)/len(A); mb=sum(B)/len(B)
cov=sum((a-ma)*(b-mb) for a,b in zip(A,B))
corr=cov/math.sqrt(sum((a-ma)**2 for a in A)*sum((b-mb)**2 for b in B))
print(f"  explicit-formula hinge: corr(psi-x, zeros) = {corr:.3f} from 60 zeros")
ok(corr>0.85,
   "the explicit formula rebuilds the primes from the zeros (the hinge)")

print(f"\nALL {checks} CHECKS PASSED")
```

Chapter 2

One Floor Up: Ramanujan's τ and Its L-Function

The opening chapter laid the abelian ground floor of the building. The wheel's Dirichlet characters were each a one-dimensional Frobenius symbol $\chi(p)$; each opened a Dirichlet L-function with its own critical line and its own copy of the wall; and together they are the wheel's abelian family. This chapter climbs one flight. Ramanujan's tau function and its L-function are the first object past the wheel's characters, and the striking thing is how little has to be relearned: the same Euler product, the same functional equation, the same critical line, the same wall—only now two-dimensional instead of one, modular instead of cyclotomic, and standing on the very hyperbolic surface where Selberg's trace formula lives. One piece of its wall has even been knocked down, by geometry, exactly where a real geometry was always the thing that could bring a wall down.

Scope

The foundation reached the abelian ground floor and stopped one dimension short. The wheel's characters were the Frobenius data of the abelian extensions of $\mathbb{Q}$—a single root of unity $\chi(p)$ recording how each prime split—and that is as far as the wheel's own arithmetic reaches. This chapter takes one famous object from just past that edge—Ramanujan's tau function $\tau(n)$ and its Dirichlet series $L(s, \Delta)$—and sets it beside the wheel to see what carries over. Almost everything does, and that is the point. The wheel's L-functions are the simplest members of an enormous family, and $L(s, \Delta)$ is the next member up: the abelian ground floor gives way to a non-abelian first floor, the Kronecker–Weber world of roots of unity gives way to the world of modular forms, and the Frobenius character $\chi(p)$ that ran the whole apparatus becomes the trace of a two-by-two matrix. The furniture is identical; only the dimension changes. We mark, as always, which resemblances are genuine structure, which are rhymes, and which would be mistakes to lean on. And we note where the wall, here, has actually fallen—because for this object, unlike for the Riemann zeta, a piece of it has.

2.1 The function, and where it comes from

Tau is not a free-standing definition; it is the list of coefficients of one specific modular form, the discriminant

$$\Delta(q) \;=\; q\prod_{n\geq 1}(1-q^n)^{24} \;=\; \sum_{n\geq 1}\tau(n)\,q^n, \qquad q=e^{2\pi i z}.$$

This Δ is a cusp form of weight 12 for the full modular group $\mathrm{SL}(2,\mathbb{Z})$, and in fact the unique normalized one—weight 12 is the lowest weight at which a cusp form for the modular group exists at all, and the space it lives in is one-dimensional. The first values are concrete and unremarkable-looking:

$$\tau(1)=1,\quad \tau(2)=-24,\quad \tau(3)=252,\quad \tau(4)=-1472,$$
$$\tau(5)=4830,\quad \tau(6)=-6048,\quad \tau(7)=-16744.$$

What is remarkable is the hidden order Ramanujan saw in them in 1916. The function is *multiplicative*: $\tau(mn)=\tau(m)\tau(n)$ whenever m and n are coprime, and at prime powers it obeys the recursion

$$\tau(p^{r+1}) \;=\; \tau(p)\,\tau(p^r) - p^{11}\,\tau(p^{r-1}).$$

Mordell proved this a year later, by an argument that became the theory of Hecke operators: Δ is a simultaneous eigenform of all of them, and the $\tau(p)$ are its eigenvalues. That single fact is the engine behind everything that follows, because multiplicativity is exactly what turns a Dirichlet series into an Euler product. Here it gives

$$L(s,\Delta) \;=\; \sum_{n\geq 1}\frac{\tau(n)}{n^s} \;=\; \prod_p\left(1-\tau(p)\,p^{-s}+p^{\,11-2s}\right)^{-1}.$$

2.2 The same furniture, one dimension higher

Set this Euler product beside the wheel's. A Dirichlet L-function of the wheel has a *degree-one* factor at each prime,

$$L(s,\chi)=\prod_p\left(1-\chi(p)\,p^{-s}\right)^{-1},$$

one prime and one root of unity. The tau L-function has a *degree-two* factor, a genuine quadratic in p^{-s}. That is the whole difference between the two floors. The wheel lives over the abelian extensions of $\mathbb{Q}$—the cyclotomic world of Kronecker and Weber, the GL(1) story. The tau function lives one rung up, in the GL(2) world of modular forms, where the symmetry behind the Euler product is no longer a character valued on the unit circle but a two-dimensional representation. This is the first step onto the Langlands ladder, and the wheel's characters were its bottom rung.

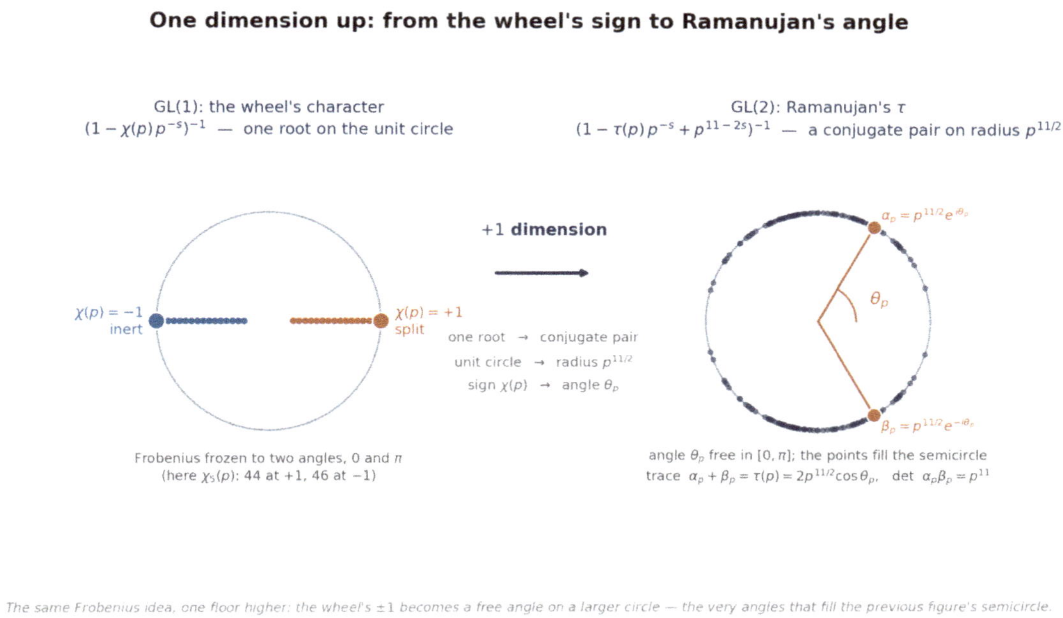

Figure 2.1: From the wheel's sign to Ramanujan's angle—the dictionary of this chapter, drawn. *Left*: the wheel's degree-one Euler factor has a single inverse root, the character value $\chi(p)$, on the unit circle; for a real character it is frozen to two angles, 0 (+1, p splits) and π (-1, p inert), and the dots are primes sorted by the conductor-5 character χ_5. *Right*: the degree-two factor of τ has a *conjugate pair* $\alpha_p, \beta_p = p^{11/2}e^{\pm i\theta_p}$ on the circle of radius $p^{11/2}$, with trace $\alpha_p+\beta_p = \tau(p) = 2p^{11/2}\cos\theta_p$ and determinant $\alpha_p\beta_p = p^{11}$ (the pair for $p = 11$, at $\theta = 60°$, drawn in red). Climbing one dimension, the wheel's ±1 unfreezes into the angle θ_p, and the two allowed points become a pair free to fill the circle—the same angles whose distribution is the semicircle of Figure 2.2.

Everything else transfers intact. Hecke proved that $L(s,\Delta)$ extends to an entire function of s and satisfies a functional equation: completing it as $\Lambda(s) = (2\pi)^{-s}\,\Gamma(s)\,L(s,\Delta)$, one has

$$\Lambda(s) = \Lambda(12-s),$$

the same reflection symmetry that organized ζ and the wheel's L-functions, now folded about the line $\mathrm{Re}(s) = 6$ instead of $\mathrm{Re}(s) = \frac{1}{2}$. (Shift by $\frac{11}{2}$ to normalize and the centre returns to $\frac{1}{2}$, and $L(s,\Delta)$ becomes a standard degree-two L-function with $s \leftrightarrow 1-s$.) And with the functional equation comes its own Riemann Hypothesis: the nontrivial zeros of $L(s,\Delta)$ are conjectured to lie on that central line. It is open, in precisely the way the wheel's Generalized Riemann Hypothesis is open. The wall did not soften by climbing a floor; it reappeared in the same place.

2.3 $\tau(p)$ is a Frobenius trace

The deepest line of the dictionary is the one that turns τ back into the language of the wheel. Attached to Δ there is a two-dimensional Galois representation—constructed by Deligne—

$$\rho:\ \mathrm{Gal}(\overline{\mathbb{Q}}/\mathbb{Q}) \longrightarrow \mathrm{GL}_2, \qquad \mathrm{tr}\,\rho(\mathrm{Frob}_p) = \tau(p), \qquad \det\rho(\mathrm{Frob}_p) = p^{11}.$$

Read that against the wheel. There, the Frobenius at p was the character value $\chi(p)$, a single number recording how p split in an abelian extension—the splitting law the wheel's characters obey in the quadratic fields. Here it is the *trace* of a two-by-two Frobenius matrix. The tau values are Frobenius data, exactly as the wheel's character values were—only promoted from one dimension to two. The same idea, the same word, a larger matrix.

This is why Ramanujan's congruences feel so much like the wheel even though they come from a modular form. The most famous,

$$\tau(n) \equiv \sigma_{11}(n) \pmod{691},$$

ties τ to the elementary divisor-power sum modulo 691—a prime that appears only because it divides the numerator of the Bernoulli number B_{12}, the same weight-12 fingerprint. Sharper still is the behaviour modulo 23: there $\tau(p)$ is governed by how p splits, vanishing precisely when p is a non-residue, so that the congruence is reading Frobenius in a field tied to $\mathbb{Q}(\sqrt{-23})$. That is the wheel's own logic—small-prime residue conditions steering arithmetic, exactly the splitting laws of the wheel's own characters—surfacing inside a degree-two object. Serre and Swinnerton-Dyer mapped out exactly which primes $(2, 3, 5, 7, 23, 691)$ carry such congruences, and why: they are the primes at which the Galois representation is smaller than it generically should be.

2.4 The geometry, and the wall that fell

Two things about Δ point straight back at the spectral picture. First, where it lives: a modular form for $\mathrm{SL}(2,\mathbb{Z})$ is a function on the quotient of the upper half-plane by the modular group—the modular surface, the hyperbolic geometry of integer Möbius transformations that carried Selberg's trace formula. The tau function is not near that surface; it is *defined on it.* Climb one floor from the wheel and you are standing on the spectral surface itself.

Second, and this is the line worth carrying away: one piece of the tau wall has actually been knocked down. Ramanujan conjectured a bound on the size of the coefficients,

$$|\tau(p)| \;\le\; 2\,p^{11/2},$$

equivalently that the two local roots α_p, β_p in $1-\tau(p)X+p^{11}X^2 = (1-\alpha_pX)(1-\beta_pX)$ satisfy $|\alpha_p| = |\beta_p| = p^{11/2}$, so that $\tau(p) = 2p^{11/2}\cos\theta_p$ for a real angle. This is the *local* Riemann Hypothesis for Δ, and it is a theorem—proved by Deligne in 1974 as a consequence of the Weil conjectures, by the same mechanism these methods keep circling: Frobenius eigenvalues on the cohomology of a variety over a finite field, forced onto a circle of the right radius by genuine geometry. The *global* Riemann Hypothesis for $L(s, \Delta)$—the location of its zeros—remains open. The split is the recurring one: where a real geometry hands over a real operator, the analogue of the wall comes down; where it does not, the wall stands. For tau, the local case fell and the global case did not.

A last rhyme, marked as a rhyme. The angles θ_p are equidistributed, with respect to the Sato–Tate measure $\frac{2}{\pi}\sin^2\theta\, d\theta$, a fact now proved for forms like Δ. It is tempting to set this beside the random-matrix statistics of the zeros, and the two are kin in spirit—both are equidistribution laws extracted from the primes. But they describe different things: Sato–Tate is a law about the coefficients $\tau(p)$, the random-matrix statistics a law about the zeros. They rhyme; they are not the same theorem. Both halves of this section are visible in a single picture (Figure 2.2): the confinement and the semicircle, the wall that fell and the rhyme.

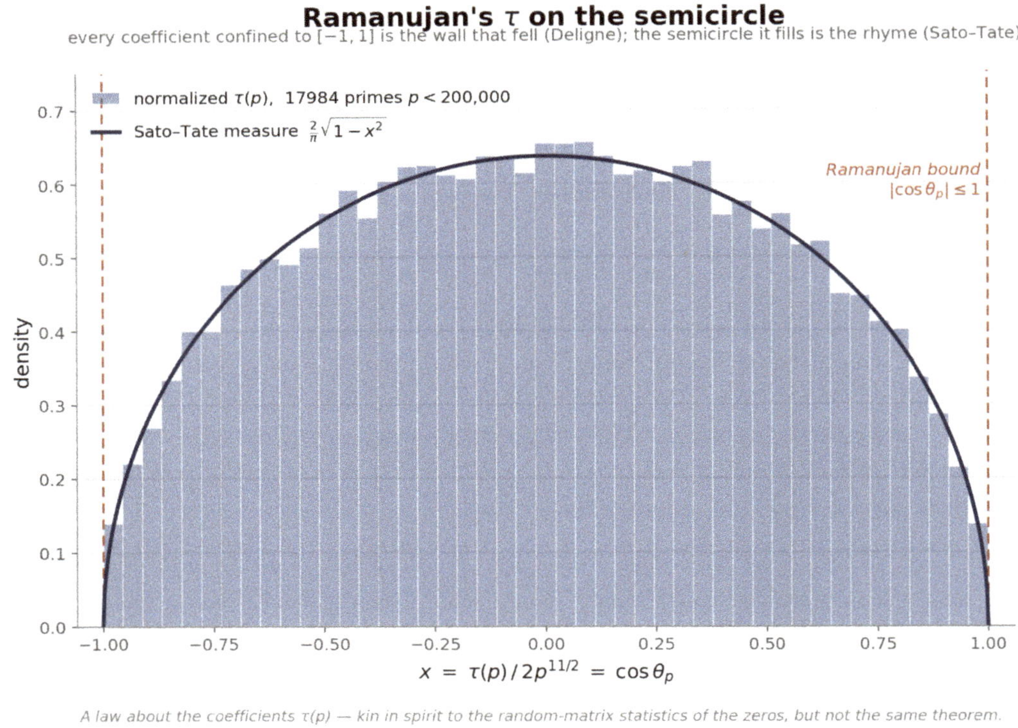

Figure 2.2: Ramanujan's τ on the semicircle. Each of the 17,984 primes $p < 200{,}000$ contributes the normalized coefficient $x_p = \tau(p)/2p^{11/2} = \cos\theta_p$; the bars are their density, the curve the Sato–Tate measure $\frac{2}{\pi}\sqrt{1-x^2}$. That every bar lies within $[-1,1]$ is the Ramanujan bound—the *local* Riemann Hypothesis, proved by Deligne from the Weil conjectures, the one piece of the wall that has actually fallen. That the bars fill the semicircle is the Sato–Tate law, a statement about the coefficients $\tau(p)$; it is kin in spirit to the random-matrix statistics of the zeros, but it is not the same theorem, and the *global* Riemann Hypothesis for $L(s,\Delta)$—the location of the zeros—is not in this picture and is not known.

What this is and is not

What is genuine here is structural and strong. The tau L-function belongs to the same family as the wheel's—Euler product, analytic continuation, functional equation, a Riemann Hypothesis—and it is the first member past the abelian ground floor, the GL(2) rung above the wheel's GL(1). The identification $\tau(p) = \operatorname{tr}\operatorname{Frob}_p$ is the wheel's Frobenius character raised one dimension, and its congruence modulo 23 is the characters' own splitting law reappearing inside a degree-two object. The form lives on the hyperbolic modular surface that carried Selberg's trace formula. And the local Riemann Hypothesis, the Ramanujan bound, is a theorem, proved by exactly the geometry these methods keep pointing at.

What this is not is a bridge across anything new. The global Riemann Hypothesis for $L(s, \Delta)$ is as open as Riemann's own, and climbing a floor did not lower it. The numerical coincidences should be left as what they are: 691 is the Bernoulli prime of weight 12, not a wheel modulus, and the resemblance to the wheel is the resemblance of two members of one family, not an arithmetic identity between them. Sato–Tate rhymes with the statistics of the zeros without being them. The honest summary is the most exciting one available: Ramanujan's τ is the wheel one full floor up, carrying every piece of furniture the foundation taught the reader to recognize and standing on the geometry the spectral chapters walk toward—and it is the doorway through which the rest of the climb continues.

Verification

The arithmetic claims below are checked against $\tau(n)$ computed directly from the q-expansion of Δ: multiplicativity, the Hecke relation at prime powers, the mod-691 congruence, the Ramanujan bound, and the mod-23 splitting law that mirrors the wheel's own characters. Standard library only.

```
#!/usr/bin/env python3
"""Verification for 'One Floor Up: Ramanujan's tau'.
tau(n) from Delta = q * prod (1-q^n)^24; standard library only."""

N = 120
P = [0] * N
P[0] = 1
for n in range(1, N):                    # multiply by prod (1-q^n)^24 mod q^N
    for _ in range(24):
        for i in range(N - 1, n - 1, -1):
            P[i] -= P[i - n]
tau = [0] + [P[k - 1] for k in range(1, N)]    # tau[k], 1 <= k <= N-1

def sigma11(n):
    return sum(d ** 11 for d in range(1, n + 1) if n % d == 0)

def is_prime(n):
    if n < 2: return False
    d = 2
    while d * d <= n:
        if n % d == 0: return False
        d += 1
    return True

def legendre(a, p):                      # (a/p) for odd prime p
    a %= p
    if a == 0: return 0
    return 1 if pow(a, (p - 1) // 2, p) == 1 else -1

C = 0
def check(name, cond):
    global C
    assert cond, "FAIL: " + name
    C += 1
    print("  PASS  " + name)

check("tau(1..5) = 1,-24,252,-1472,4830",
      tau[1:6] == [1, -24, 252, -1472, 4830])
```

```
check("tau(6) = tau(2)tau(3)",  tau[6] == tau[2] * tau[3])
check("tau(10) = tau(2)tau(5)", tau[10] == tau[2] * tau[5])
check("tau(15) = tau(3)tau(5)", tau[15] == tau[3] * tau[5])

check("tau(4) = tau(2)^2 - 2^11", tau[4] == tau[2] ** 2 - 2 ** 11)
check("tau(9) = tau(3)^2 - 3^11", tau[9] == tau[3] ** 2 - 3 ** 11)
check("tau(25) = tau(5)^2 - 5^11", tau[25] == tau[5] ** 2 - 5 ** 11)

check("tau(n) = sigma_11(n) mod 691, n<=6",
      all((tau[n] - sigma11(n)) % 691 == 0 for n in range(1, 7)))

check("|tau(p)| <= 2 p^(11/2), p<=13",
      all(abs(tau[p]) <= 2 * p ** 5.5 for p in [2, 3, 5, 7, 11, 13]))

# Frobenius mod 23 (Serre, Swinnerton-Dyer): for p != 23,
# tau(p) == 0 (mod 23)  iff  p is a non-residue mod 23  iff  p does not split
# in Q(sqrt -23) -- the wheel's splitting law, surfacing inside a degree-2 object.
check("tau(p) = 0 mod 23  iff  (p/23) = -1,  for p < %d, p != 23" % N,
      all((tau[p] % 23 == 0) == (legendre(p, 23) == -1)
          for p in range(2, N) if is_prime(p) and p != 23))

print("\n%d/%d PASS" % (C, C))
```

Part II

The Five Faces of the Hypothesis

Chapter 3

The Oscillation Face

Here is the first face of the wall, and the most concrete. In this tongue the Riemann Hypothesis is the claim that a sum built from the primes cancels down to square-root size. Four famous sums say it—the error in the prime number theorem, Chebyshev's ψ, Mertens's M, Liouville's L—and they are not four problems but one, in four costumes, because the explicit formula turns each of them into a sum over the zeros. The lure is that these sums are utterly concrete: you can compute them, watch them wander, and find them small in every case anyone has ever reached. The turn comes in two parts. The cancellation one would have to prove is *the location of the zeros, so the face is circular; and two of its most natural sharpenings—Mertens's and Pólya's—are simply false, while holding true as far as any computer has gone. This is the chapter where one learns, once and for all, that agreement as far as the eye can see is not the same thing as truth.*

Scope

This face gathers the reformulations of the Hypothesis as statements of cancellation. Each is an equivalence in the strict sense—it holds if and only if every nontrivial zero lies on the critical line—and each is built directly on the explicit formula of the opening chapter. We state the four headline sums with their equations and their sharp forms, show in one stroke why they are the same statement, follow the lure to where it is most seductive, and then deliver the turn twice: the structural reason the face cannot be crossed, and the two celebrated conjectures that died on it. The smaller members of the family—the Riesz and Franel–Landau criteria, the Redheffer matrix—belong to the same face and are catalogued in the appendix; they add nothing to the lesson the four headliners already teach.

3.1 The four sums

Begin with the prime number theorem. Write $\pi(x)$ for the number of primes up to x and $\mathrm{Li}(x) = \int_2^x dt/\log t$ for the logarithmic integral. The theorem says $\pi(x) \sim \mathrm{Li}(x)$;

the Hypothesis says exactly how close:

$$\pi(x) = \mathrm{Li}(x) + O(x^{1/2+\varepsilon}) \quad \Longleftrightarrow \quad \text{RH.} \tag{3.1}$$

Schoenfeld made the bound explicit: on RH, $|\pi(x) - \mathrm{Li}(x)| < \frac{1}{8\pi}\sqrt{x}\log x$ for $x \geq 2657$. The same statement reads more cleanly through Chebyshev's weighted count $\psi(x) = \sum_{p^k \leq x} \log p$, where it becomes

$$\psi(x) = x + O(\sqrt{x}\,(\log x)^2) \quad \Longleftrightarrow \quad \text{RH.} \tag{3.2}$$

Two more sums say the same thing in the language of multiplicative functions. The Möbius function $\mu(n)$ is 0 if n has a squared factor and $(-1)^{(\text{number of prime factors})}$ otherwise; its Dirichlet series is $1/\zeta(s) = \sum_n \mu(n)n^{-s}$. The Liouville function $\lambda(n) = (-1)^{\Omega(n)}$ counts prime factors with multiplicity, with $\zeta(2s)/\zeta(s) = \sum_n \lambda(n)n^{-s}$. Their summatory functions are

$$M(x) = \sum_{n \leq x} \mu(n), \qquad L(x) = \sum_{n \leq x} \lambda(n), \tag{3.3}$$

and for each, square-root cancellation is again the Hypothesis:

$$M(x) = O(x^{1/2+\varepsilon}) \Longleftrightarrow \text{RH}, \qquad L(x) = O(x^{1/2+\varepsilon}) \Longleftrightarrow \text{RH.} \tag{3.4}$$

Figure 3.1 shows all four. They look like four different oscillations, but the eye should be trained to see one: each is a prime-built sum wobbling around its mean at the scale of $\sqrt{x}$. The verification confirms that to 10^6 all four stay well inside an $x^{0.6}$ band ($\max|\pi - \mathrm{Li}| = 135$, $\max|\psi - x| = 511$, $\max|M| = 368$, $\max|L| = 1253$).

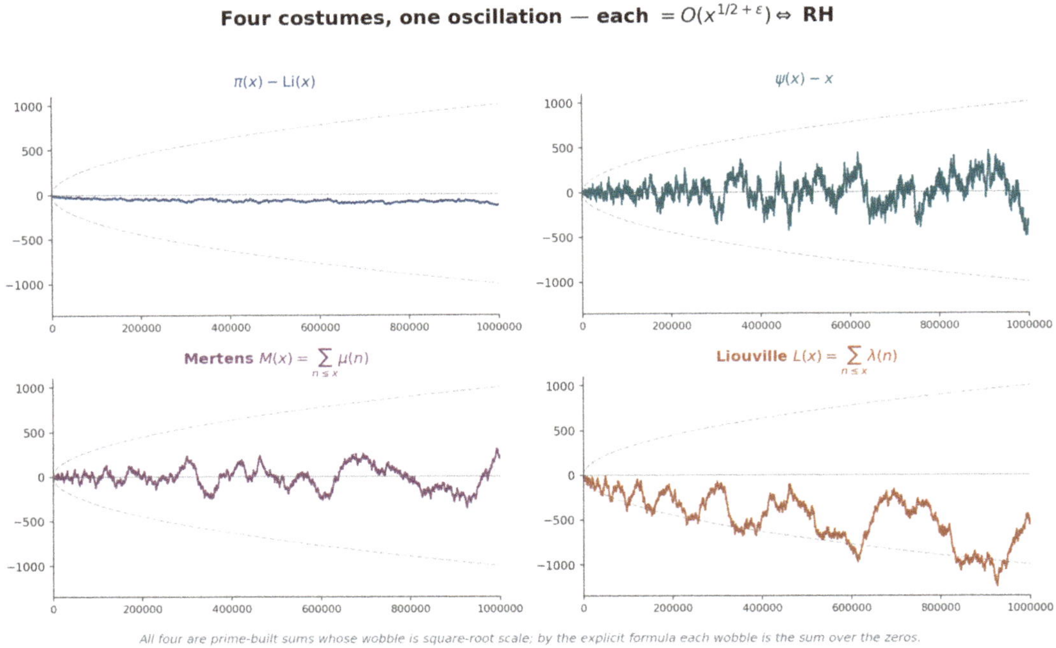

Figure 3.1: Four costumes, one oscillation. The prime number theorem's error $\pi(x) - \mathrm{Li}(x)$, Chebyshev's $\psi(x) - x$, the Mertens sum $M(x)$, and the Liouville sum $L(x)$, each against the $\pm\sqrt{x}$ band. All four are prime-built sums wobbling at square-root scale, and each being $O(x^{1/2+\varepsilon})$ is equivalent to the Hypothesis. Liouville's L pushes hardest against the band—a foreshadowing.

3.2 Why they are one

The reason the four are a single statement is the explicit formula. For ψ it is exact:

$$\psi(x) = x - \sum_{\rho} \frac{x^{\rho}}{\rho} - \log(2\pi) - \tfrac{1}{2}\log(1 - x^{-2}), \tag{3.5}$$

the sum over the nontrivial zeros ρ. The wobble $\psi(x) - x$ is nothing but $-\sum_{\rho} x^{\rho}/\rho$. Now the size of each term is set by the real part of its zero, $|x^{\rho}| = x^{\mathrm{Re}\,\rho}$, so a zero on the critical line contributes at scale $\sqrt{x}$ and a zero off the line contributes at scale $x^{\mathrm{Re}\,\rho}$, larger. Square-root cancellation is therefore not a separate fact to be proved; it is the statement that no zero has real part exceeding $\frac{1}{2}$, written as a bound on a sum. The Mertens and Liouville sums have their own explicit formulas, through $1/\zeta$ and $\zeta(2s)/\zeta(s)$, with the same consequence: the zeros of ζ are the frequencies of all four oscillations, and the real parts of the zeros are their amplitudes.

This is what makes the face one face. The verification shows it directly: feed the first sixty zeros into Eq. (3.5) and the recovered $\psi(x) - x$ tracks the true one at correlation 0.92. The oscillation is the zeros, seen from the prime side.

3.3 The lure

The pull of this face is the pull of the concrete. These are not abstract functionals; they are integer-valued sums you can tabulate. Watch $M(x)$ for a while and it drifts a few hundred either way, never far from zero; $L(x)$ leans gently negative; $\pi(x) - \mathrm{Li}(x)$ barely stirs. Everything stays small, and stays small no matter how far you push—to a million, a billion, a trillion. It is almost impossible, sitting in front of such a table, not to feel that the bound is simply *true*, and that a clever enough estimate of the sum would prove it and topple the wall. The whole face is the temptation to believe that what is small in every case you can see is small in every case there is.

3.4 The turn, and why

The first turn is structural, and it is the turn this whole book keeps meeting. To prove that $\psi(x) - x$ is $O(\sqrt{x}\,(\log x)^2)$ you must bound $\sum_\rho x^\rho/\rho$, and to bound that you must know the real parts of the zeros do not exceed $\frac{1}{2}$—which is the Hypothesis. The sum you are trying to estimate is the zeros; the estimate you need is their location. The lever and the destination are the same object, and no amount of cleverness on the prime side reaches around to the zeros, because the explicit formula only runs one way that we can use: it expresses the primes through the zeros, never the zeros through anything we can independently hold.

3.5 The lures that dissolved

The second turn is sharper, and it is unique to this face. Strip the ε from the bound and you get conjectures that are cleaner, stronger, and false—and they are false in the cruellest way, by holding true across every case anyone can compute.

Mertens conjectured the tidy bound $|M(x)| \leq \sqrt{x}$ for all $x > 1$. It holds throughout the verification's range and far beyond; and it is false. Odlyzko and te Riele proved in 1985 that $\limsup M(x)/\sqrt{x} > 1.06$ and $\liminf < -1.009$, so the bound is broken infinitely often—yet their proof is non-constructive, and to this day no explicit x where it fails is known, the smallest believed to be astronomically large. Pólya conjectured that $L(x) \leq 0$ for every $x \geq 2$: that among the integers up to any bound, those with an odd number of prime factors are never outnumbered. It holds to $x = 10^6$, where the verification confirms it, and well past; and it is false, the first x with $L(x) > 0$ being 906,150,257. Figure 3.2 draws both: each conjecture sitting comfortably inside its forbidden line across the whole computable plot, each destined to leave it at a horizon no plot can reach.

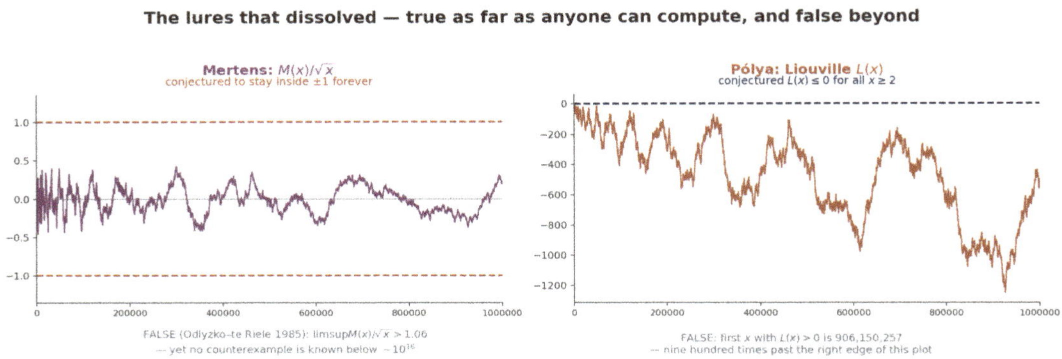

Figure 3.2: The lures that dissolved. *Left:* $M(x)/\sqrt{x}$ stays inside ± 1—the Mertens conjecture—throughout, and is false (Odlyzko–te Riele, $\limsup > 1.06$), with no counterexample known below $\sim 10^{16}$. *Right:* $L(x) \leq 0$—Pólya's conjecture—holds across the plot and is false, first broken at 906,150,257. True as far as anyone can compute; false beyond.

The lesson is exact and it is why the equivalences keep their ε. The true statement, $M(x) = O(x^{1/2+\varepsilon})$, survives because it asks only for square-root cancellation up to an arbitrarily small slack. The false sharpening, $|M(x)| \leq \sqrt{x}$, asks for the slack to be exactly zero, and that is more than the zeros permit: clustering many zeros just above the critical line lets the sum push a hair past $\sqrt{x}$, infinitely often, in a way no finite computation will ever display. The gap between the surviving equivalence and the dead conjecture is the whole difference between a theorem and a coincidence, and this face is where one learns to feel it.

What this is and is not

What is genuine here is a tight cluster of equivalences and the single mechanism beneath them. The four sums—$\pi - \mathrm{Li}$, $\psi - x$, M, L—are each, in sharp and classical form, equivalent to the Hypothesis, and the explicit formula shows why they are one statement: the oscillation is the zeros, its amplitude their real parts. The verification exhibits all of it—the four staying square-root small, the reconstruction from sixty zeros, the two conjectures holding in range.

What it is not is a way to bound the sum. The cancellation one would prove is the location one is trying to find, and the face is closed for that reason alone. And it carries the survey's bluntest warning, paid in two famous deaths: a pattern can hold across every number a machine will ever reach and still be false. Mertens's $\sqrt{x}$ and Pólya's sign rule were not careless guesses; they were the natural, beautiful, slightly-too- strong readings of a true equivalence, and the primes broke both. The oscillation face is honest about its own lure, and the honesty is the only defence it has.

Verification

The script below, standard library only and deterministic, computes the four sums to 10^6 and confirms they stay $O(x^{0.6})$; checks that the Mertens and Pólya conjectures hold throughout that range while recording that both are false; and rebuilds $\psi(x) - x$ from the first sixty zeros to show the oscillation is the zeros.

```
#!/usr/bin/env python3
"""Verification for 'The Oscillation Face'. Standard library only; deterministic."""
from math import log, isqrt, sqrt, cos, sin, pi as PI
import array, time
t0=time.time()
N=10**6
checks=0
def ok(cond,msg):
    global checks
    assert cond,"FAIL: "+msg
    checks+=1; print("PASS:",msg)

# smallest-prime-factor sieve -> primes, mu, Omega-parity, von Mangoldt Lambda
spf=list(range(N+1))
i=2
while i*i<=N:
    if spf[i]==i:
        for j in range(i*i,N+1,i):
            if spf[j]==j: spf[j]=i
    i+=1
isprime=bytearray(N+1)
for n in range(2,N+1):
    if spf[n]==n: isprime[n]=1
mu=array.array('b',[0])*(N+1); mu[1]=1
parity=array.array('b',[0])*(N+1)       # Omega(n) mod 2
Lam=array.array('d',[0.0])*(N+1)        # von Mangoldt
for n in range(2,N+1):
    x=n; om=0; sq=True; first=spf[n]; powerof=True
    while x>1:
        p=spf[x]; c=0
        while x%p==0: x//=p; c+=1; om+=1
        if c>1: sq=False
        if p!=first: powerof=False
    parity[n]=om%2
    mu[n]=0 if not sq else (1 if om%2==0 else -1)
    if powerof: Lam[n]=log(first)       # n is a prime power

# single pass: accumulate pi, psi, Mertens M, Liouville L, and Li (trapezoid)
pi=0; psi=0.0; M=1; L=1; li=0.0   # n=1: mu(1)=1, lambda(1)=1
mx_piLi=mx_psi=mx_M=mx_L=0.0
mert_ok=True; polya_ok=True
inv_prev=1.0/log(2)
for x in range(2,N+1):
    if isprime[x]: pi+=1
    psi+=Lam[x]; M+=mu[x]; L+= 1 if parity[x]==0 else -1
    inv=1.0/log(x)
    if x>2: li+=0.5*(inv_prev+inv)      # trapezoid for Li(x)=int_2^x dt/ln t
    inv_prev=inv
    r=sqrt(x)
    if abs(pi-li)>mx_piLi: mx_piLi=abs(pi-li)
    if abs(psi-x)>mx_psi: mx_psi=abs(psi-x)
    if abs(M)>mx_M: mx_M=abs(M)
    if abs(L)>mx_L: mx_L=abs(L)
    if abs(M)>r: mert_ok=False           # Mertens conjecture |M(x)|<=sqrt(x)
    if x>=2 and L>0: polya_ok=False      # Polya conjecture L(x)<=0
```

```
e=N**0.6
print(f"  to 10^6 (x^0.6={e:.0f}):  max|pi-Li|={mx_piLi:.0f}  max|psi-x|={mx_psi:.0f}"
      f"  max|M|={mx_M:.0f}  max|L|={mx_L:.0f}")
ok(mx_piLi<e and mx_psi<e and mx_M<e and mx_L<e,
   "all four (pi-Li, psi-x, M, L) stay O(x^0.6) to 10^6: one sqrt-scale oscillation")

ok(mert_ok,
   "Mertens |M(x)|<=sqrt(x) holds to 10^6 -- yet FALSE (Odlyzko-te Riele 1985)")
ok(polya_ok,
   "Polya L(x)<=0 holds to 10^6 -- yet FALSE (first sign change at 906150257)")

# the mechanism: the oscillation IS the zeros (explicit formula rebuilds psi(x)-x)
GAMMA=[14.13472514,21.02203964,25.01085758,30.42487613,32.93506159,37.58617816,
 40.91871901,43.32707328,48.00515088,49.77383248,52.97032148,56.4462477,59.347044,
 60.83177852,65.11254405,67.07981053,69.54640171,72.06715767,75.7046907,77.14484007,
 79.33737502,82.91038085,84.73549298,87.42527461,88.80911121,92.49189927,94.65134404,
 95.87063423,98.83119422,101.31785101,103.72553804,105.44662305,107.16861118,
 111.02953554,111.87465918,114.32022092,116.22668032,118.79078287,121.370125,
 122.94682929,124.25681855,127.51668388,129.5787042,131.08768853,133.4977372,
 134.75650975,138.11604205,139.73620895,141.1237074,143.11184581,146.00098249,
 147.42276534,150.05352042,150.92525761,153.02469381,156.11290929,157.59759182,
 158.84998817,161.18896414,163.03070969]
def psi_zeros(t):
    Lx=log(t); s=0.0
    for g in GAMMA: s+=(0.5*cos(g*Lx)+g*sin(g*Lx))/(0.25+g*g)
    return t-log(2*PI)-0.5*log(1-t**-2)-2*sqrt(t)*s
# actual psi on a grid from the Lambda we already have
def psi_at(t):
    it=int(t); return sum(Lam[k] for k in range(2,it+1))
xs=[2+0.2*k for k in range(1,490)]
A=[psi_at(t)-t for t in xs]; B=[psi_zeros(t)-t for t in xs]
ma=sum(A)/len(A); mb=sum(B)/len(B)
cov=sum((a-ma)*(b-mb) for a,b in zip(A,B))
corr=cov/sqrt(sum((a-ma)**2 for a in A)*sum((b-mb)**2 for b in B))
print(f"  oscillation IS the zeros: corr(psi-x, zeros) = {corr:.3f}")
ok(corr>0.85, "the oscillation IS the zeros: the explicit formula rebuilds psi(x)-x")

print(f"\nALL {checks} CHECKS PASSED   ({time.time()-t0:.1f}s)")
```

Chapter 4

The Positivity Face

On this face the wall becomes a sign. The Riemann Hypothesis turns into the claim that a particular sequence of real numbers never dips below zero—Li's coefficients λ_n—and, one level deeper, that a quadratic form assembled from the explicit formula is never negative, which is Weil's criterion and the quiet engine of the whole spectral programme. The lure is as clean as the survey offers: you can compute the λ_n, watch them come out positive and rising, and feel certain that a positivity visible in every case must hold in all. The turn is geometric and exact. Each λ_n is a sum over the zeros, and the map $\rho \mapsto 1 - 1/\rho$ carries the critical line precisely onto the unit circle, so positivity holds exactly when every zero's image lands on that circle. Let one zero slip off the line and its image leaves the circle, the powers $(1 - 1/\rho)^n$ swell, and some λ_n is dragged below zero. Positivity is not a lever on the zeros; it is a hostage to them.

Scope

This face collects the reformulations of the Hypothesis as positivity conditions. We take the two that matter: Li's criterion, which is fully explicit and computable, and Weil's, which is the form the condition takes inside the explicit formula and which the operator approaches all aim at. For Li's criterion we give both definitions of λ_n, prove the small piece of geometry that makes the whole thing transparent—that the critical line is the unit circle—and then run the lure and the turn against it directly, with sixty zeros. For Weil's criterion we state the positivity and say plainly what it would take to use it, and why no one can. As always we mark what is proved, what is equivalent, and what is merely hoped.

4.1 Li's criterion

In 1997 Xian-Jin Li attached to the zeta function a sequence of real numbers. Write $\xi(s)$ for the completed function of the opening chapter, $\xi(s) = \frac{1}{2}s(s-1)\pi^{-s/2}\Gamma(s/2)\zeta(s)$, entire and symmetric under $s \mapsto 1 - s$. Li's coefficients are its logarithmic Taylor data at $s = 1$,

$$\lambda_n = \frac{1}{(n-1)!}\,\frac{d^n}{ds^n}\left[s^{n-1}\log\xi(s)\right]_{s=1}, \qquad n = 1, 2, 3, \ldots \tag{4.1}$$

That definition is opaque, but it has an exact rewriting as a sum over the nontrivial zeros, and this is the form to hold onto:

$$\lambda_n = \sum_{\rho}\left[1 - \left(1 - \frac{1}{\rho}\right)^n\right], \tag{4.2}$$

the sum taken over zeros in conjugate pairs so that each λ_n is real. Li's theorem is the equivalence

$$\lambda_n \geq 0 \text{ for every } n \geq 1 \qquad \Longleftrightarrow \qquad \text{RH}. \tag{4.3}$$

The Hypothesis has become a single word—*non-negative*—applied to a list of numbers one can actually compute. Nothing about $\lambda_n \geq 0$ looks like a statement about the location of zeros. That it is one is the content of the next section.

4.2 The critical line is the unit circle

Look at the quantity being raised to the nth power in Eq. (4.2), the Möbius transformation

$$w = 1 - \frac{1}{\rho} = \frac{\rho - 1}{\rho}. \tag{4.4}$$

Put a zero on the critical line, $\rho = \frac{1}{2} + i\gamma$, and compute the size of its image. With $|\rho|^2 = \frac{1}{4} + \gamma^2$,

$$\left|1 - \tfrac{1}{\rho}\right|^2 = \frac{|\rho - 1|^2}{|\rho|^2} = \frac{(\frac{1}{2})^2 + \gamma^2}{(\frac{1}{2})^2 + \gamma^2} = 1. \tag{4.5}$$

Every zero on the line has $|w| = 1$: the map sends the critical line exactly onto the unit circle. The two halves of the strip go to the two sides of the circle—a zero with $\operatorname{Re}\rho > \frac{1}{2}$ maps to the inside, $|w| < 1$, and one with $\operatorname{Re}\rho < \frac{1}{2}$ maps to the outside, $|w| > 1$. The verification confirms the identity to machine zero for all sixty test zeros. Figure 4.1 draws the whole picture: the critical strip on the left, the unit circle on the right, the on-line zeros landing on the circle and a hypothetical off-line pair landing off it.

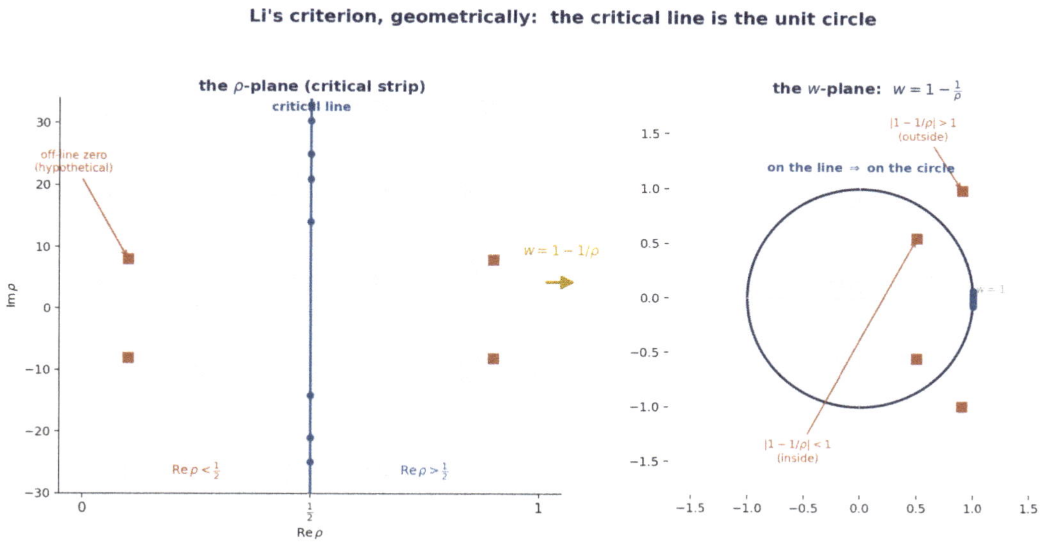

Figure 4.1: Li's criterion, geometrically. The Möbius map $w = 1 - 1/\rho$ carries the critical strip to the w-plane and the critical line $\operatorname{Re}\rho = \frac{1}{2}$ onto the unit circle. On-line zeros (blue) land on the circle, clustering near $w = 1$ as the height grows; a zero off the line (red) lands off the circle, inside if $\operatorname{Re}\rho > \frac{1}{2}$ and outside if $\operatorname{Re}\rho < \frac{1}{2}$. Li's coefficients stay non-negative exactly when every image is on the circle.

Now the criterion reads itself. Each $\lambda_n = \sum_\rho (1 - w_\rho^n)$. When a zero is on the line, $|w_\rho| = 1$, so w_ρ^n rides around the unit circle without growing and its paired contribution stays bounded and, summed correctly, non-negative. The positivity of every λ_n is precisely the statement that no w_ρ ever has modulus greater than one—that no zero lies to the left of the critical line—which, with the functional equation's mirror, is the Hypothesis.

4.3 The lure

Compute. Feeding the first sixty zeros into Eq. (4.2) gives

$$\lambda_1, \ldots, \lambda_{10} = 0.019,\ 0.076,\ 0.170,\ 0.303,\ 0.472,\ 0.679,\ 0.922,\ 1.201,\ 1.516,\ 1.866, \tag{4.6}$$

every one positive, and rising smoothly (the blue curve of Figure 4.2). The verification confirms it. This is the pull of the face: the numbers are concrete, they are increasing, and they stay positive for as many as one cares to compute—thousands have been checked, all positive. It is hard not to read the trend as a law and conclude that $\lambda_n > 0$ forever, which would be the Hypothesis proved.

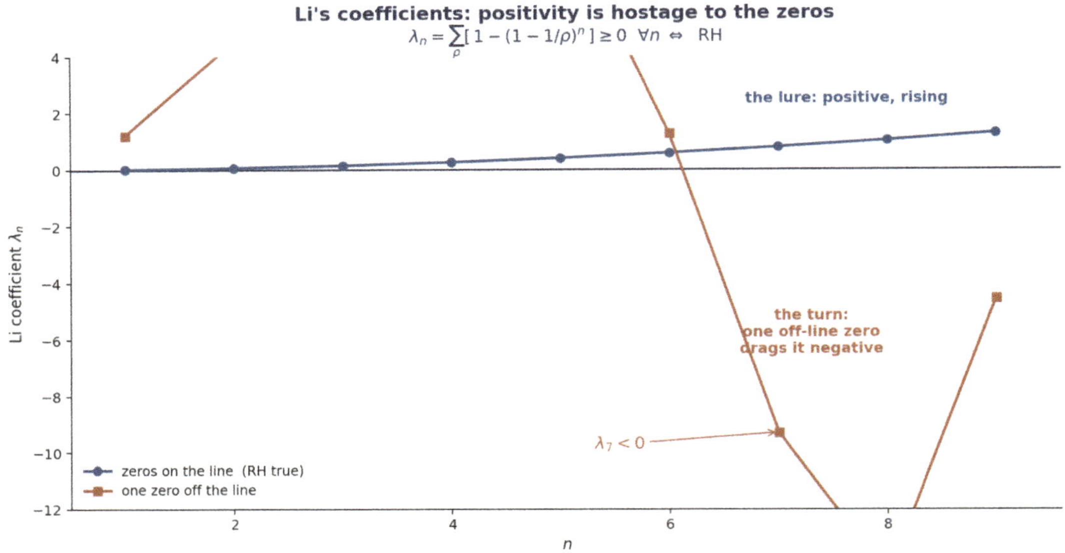

Figure 4.2: Positivity is hostage to the zeros. With every zero on the line (blue) the Li coefficients are positive and rising—the lure. Introduce a single zero off the line and the sequence (red) is dragged below zero, here first at $n = 7$: $(1 - 1/\rho)^n$ grows once $|1 - 1/\rho| > 1$, and it grows without bound. Positivity for all n is exactly the Hypothesis.

4.4 The turn, and why

Suppose a single zero sat off the line. By the functional equation it comes with its mirror, and one of the pair has real part below $\frac{1}{2}$, hence image w with $|w| > 1$. Then w^n does not ride the circle; it spirals outward, its modulus growing like $|w|^n$, and the term $1 - w^n$ in Eq. (4.2) swings to a large negative real part whenever the phase points the right way. No finite collection of well-behaved on-line zeros can outweigh an exponentially growing term forever, and so some λ_n must go negative. The verification shows it with one hypothetical off-line zero at $\mathrm{Re}\,\rho = 0.1$, where $|1 - 1/\rho| = 1.339$: the sequence stays positive through $n = 6$ and then breaks, $\lambda_7 < 0$.

That is the equivalence made mechanical, and it is also the turn. To prove $\lambda_n \geq 0$ for all n one must rule out any w with modulus exceeding one—which is to say, locate every zero on the line. So the positivity one is asked to establish is not a fact *about* the zeros that might be reached from elsewhere; it is their position rewritten as a sign. Each new positive λ_n is one more zero confirmed on the line, and not one step nearer a reason they all must be: the confidence grows and the proof does not, exactly as it did not while the Mertens sum stayed small.

4.5 Weil's positivity

Beneath Li's criterion sits the positivity the spectral approaches actually chase. Weil cast the explicit formula as a bilinear identity. For a suitable test function g with transform $\widehat{g}$, the sum over the zeros equals an archimedean term plus a sum over the primes,

$$\sum_{\rho} \widehat{g}(\rho) = \widehat{g}(0) + \widehat{g}(1) - \sum_{p,\,k} \frac{\log p}{p^{k/2}}\, g(k \log p) - (\text{archimedean}), \tag{4.7}$$

and from it one builds a Hermitian form $W(g, g)$ by pairing g with itself. Weil's criterion is the equivalence

$$W(g, g) \geq 0 \;\; \text{for all admissible } g \qquad \Longleftrightarrow \qquad \text{RH}. \tag{4.8}$$

The reason is the same geometry one floor down: each zero contributes a term whose sign is governed by whether ρ sits on the critical line, and the form is a sum of squares exactly when they all do. This is the positivity Connes reorganizes into a trace formula, and the one a Hilbert–Pólya operator would supply for free: if the zeros were the eigenvalues of a self-adjoint operator, W would be the manifest norm-squared on its eigenspaces, and positivity—and the Hypothesis—would follow. The turn is that no such operator, and no independent proof that W is a square, is known. Weil's positivity is the cleanest statement of what the spectral programme must establish; it is equivalent to the wall, not a passage through it. The neighbouring-walls chapter returns to the operator dream from this exact spot.

What this is and is not

What is genuine here is a pair of exact equivalences and a piece of geometry that makes the first of them transparent. Li's criterion converts the Hypothesis into the non-negativity of a computable real sequence; the Möbius map $w = 1 - 1/\rho$ shows why, by turning "on the critical line" into "on the unit circle"; and Weil's criterion lifts the same positivity into the explicit formula, where the operator approaches must meet it. The verification exhibits all three moving parts: the line mapping to the circle to machine precision, the sixty-zero λ_n positive and rising, and a single off-line zero breaking positivity at $n = 7$.

What it is not is a way to establish the positivity. Both criteria say the sign condition is equivalent to the location of the zeros, so proving the sign requires knowing the location, which is the wall. The face is more seductive than most—a list of positive numbers is about as concrete as an open problem gets—and that is precisely its lesson: one may watch a sequence stay positive for all the n a machine will ever reach and still have proved nothing, because the one term that could turn it negative is the one off-line zero no computation can exhibit.

Verification

The script below, standard library only and deterministic, confirms the geometry and the criterion: the map $\rho \mapsto 1 - 1/\rho$ sends every on-line zero onto the unit circle ($|1 - 1/\rho| = 1$ to machine zero); Li's λ_n for $n = 1, \ldots, 10$, built from sixty zeros, are positive and increasing; and a single hypothetical off-line zero, whose image has modulus $1.339 > 1$, drives λ_n negative (first at $n = 7$).

```
#!/usr/bin/env python3
"""Verification for 'The Positivity Face'. Standard library only; deterministic."""
GAMMA=[14.13472514,21.02203964,25.01085758,30.42487613,32.93506159,37.58617816,
 40.91871901,43.32707328,48.00515088,49.77383248,52.97032148,56.4462477,59.347044,
 60.83177852,65.11254405,67.07981053,69.54640171,72.06715767,75.7046907,77.14484007,
 79.33737502,82.91038085,84.73549298,87.42527461,88.80911121,92.49189927,94.65134404,
 95.87063423,98.83119422,101.31785101,103.72553804,105.44662305,107.16861118,
 111.02953554,111.87465918,114.32022092,116.22668032,118.79078287,121.370125,
 122.94682929,124.25681855,127.51668388,129.5787042,131.08768853,133.4977372,
 134.75650975,138.11604205,139.73620895,141.1237074,143.11184581,146.00098249,
 147.42276534,150.05352042,150.92525761,153.02469381,156.11290929,157.59759182,
 158.84998817,161.18896414,163.03070969]
checks=0
def ok(cond,msg):
    global checks
    assert cond,"FAIL: "+msg
    checks+=1; print("PASS:",msg)

# ===== mechanism: w = 1 - 1/rho sends the critical line to the unit circle =====
maxdev=0.0
for g in GAMMA:
    rho=complex(0.5,g); w=1-1/rho
    maxdev=max(maxdev,abs(abs(w)-1.0))
print(f"  on-line zeros: max | |1-1/rho| - 1 | = {maxdev:.2e}")
ok(maxdev<1e-12,
   "the map rho->1-1/rho carries on-line zeros onto the unit circle (|1-1/rho|=1)")

# ===== Li's lambda_n from the on-line zeros: positive and increasing (the lure) =====
def lam_online(n):
    tot=0.0
    for g in GAMMA:
        for rho in (complex(0.5,g),complex(0.5,-g)):
            tot += (1-(1-1/rho)**n).real
    return tot
lam=[lam_online(n) for n in range(1,11)]
print("  lambda_1..10 (on-line):", [round(v,4) for v in lam])
ok(all(v>0 for v in lam) and all(lam[i]<lam[i+1] for i in range(9)),
   "Li's lambda_n > 0 and increasing for n=1..10 (positivity holds on the line)")

# ===== the turn: one hypothetical off-line zero drags positivity negative =====
off=[complex(0.1,1.0),complex(0.1,-1.0),complex(0.9,1.0),complex(0.9,-1.0)]
woff=abs(1-1/off[0])
def lam_with_off(n):
    tot=lam_online(n)
    for rho in off: tot+=(1-(1-1/rho)**n).real
    return tot
neg_at=next((n for n in range(1,21) if lam_with_off(n)<0),None)
print(f"  off-line zero Re=0.1: |1-1/rho|={woff:.3f}>1;"
      f" lambda_n<0 first at n={neg_at}")
ok(neg_at is not None,
   "one off-line zero (|1-1/rho|>1) drives some lambda_n negative: positivity <=> RH")

print(f"\nALL {checks} CHECKS PASSED")
```

Chapter 5

The Completeness Face

On this face the wall becomes a question of richness: is a certain family of functions large enough to fill a space? Nyman and Beurling found that it is exactly the Riemann Hypothesis. Take the fractional-part function, dilate it every which way, and ask whether the constant function 1 *lies in the closure of what you can build from the pieces, inside the Hilbert space* $L^2(0,1)$. *The answer is yes if and only if every zero of zeta sits on the critical line; Báez-Duarte later showed the integer dilations alone suffice. The lure is that one can compute the best approximation of* 1 *from finitely many dilations and watch the distance shrink—*1 *is visibly being closed in upon. The turn is that the dilations carry zeta inside their Mellin transform, so the family reaches downward only as far as the rightmost zero; completeness is the statement that the rightmost zero sits on the line. The distance shrinks, and proving it reaches zero is proving the wall—right down to the rate at which it shrinks, which the zeros set.*

Scope

This face is the most analytic in the survey, and the most surprising: the Hypothesis as a spanning condition in a function space. We give the Nyman–Beurling criterion and its Báez-Duarte discretization with their equations, build the bridge that makes the mechanism visible—that the fractional part is zeta in disguise—and then run the lure and the turn. The lure is a genuine, computable shrinking distance; the turn is that the distance, and even its rate of decrease, are governed by the zeros. We mark throughout what is an exact equivalence, what is conjectural about the rate, and where the approximation's honesty (it converges in mean, not pointwise) must be stated plainly.

5.1 The Nyman–Beurling criterion

Work in $L^2(0,1)$, the square-integrable functions on the unit interval, with $\langle f,g\rangle = \int_0^1 f\,g\,dt$ and $\|1\|^2 = \int_0^1 dt = 1$. For each θ with $0<\theta\le 1$ define the dilated fractional part

$$\rho_\theta(t) = \left\{\frac{\theta}{t}\right\}, \qquad 0 < t < 1, \tag{5.1}$$

where $\{x\}$ is the fractional part. These are bounded, hence in $L^2(0,1)$. Nyman and Beurling proved, around 1950, the equivalence

$$\mathbf{1} \in \overline{\operatorname{span}}\,\{\rho_\theta : 0 < \theta \le 1\} \text{ in } L^2(0,1) \qquad \Longleftrightarrow \qquad \text{RH}, \tag{5.2}$$

that is, the constant function 1 can be approximated in mean square, to any accuracy, by finite combinations $\sum_k c_k \rho_{\theta_k}$ exactly when the Hypothesis holds. (Equivalently: the closed span is all of $L^2(0,1)$.) In 2003 Báez-Duarte sharpened the family: it is enough to use the *integer* dilations $\theta = 1/n$, $n = 2, 3, 4, \ldots$, and indeed the criterion can be reduced to the convergence of a single explicit sequence of vectors. The Hypothesis has become a completeness theorem.

5.2 The bridge: the fractional part is zeta

Why a question about fractional parts should be a question about zeros is settled by one classical identity. For $\operatorname{Re} s > 0$, $s \neq 1$,

$$\zeta(s) = \frac{s}{s-1} - s \int_1^\infty \{x\}\, x^{-s-1}\, dx, \tag{5.3}$$

the fractional part generating zeta directly. The verification confirms it to eleven digits at $s = 2, 3, 4$, recovering $\pi^2/6$, Apéry's constant, and $\pi^4/90$. Feed this into the Mellin transform of the building block. Substituting $u = 1/t$,

$$\int_0^1 \left\{\frac{1}{t}\right\} t^{s-1}\, dt = \int_1^\infty \{u\}\, u^{-s-1}\, du = \frac{1}{s-1} - \frac{\zeta(s)}{s}, \qquad 0 < \operatorname{Re} s < 1, \tag{5.4}$$

and a dilation by θ multiplies this by θ^s. So every ρ_θ carries the factor $\zeta(s)/s$ in its transform, and a finite combination $\sum_k c_k \rho_{\theta_k}$ has Mellin transform

$$\left(\textstyle\sum_k c_k \theta_k^s\right)\left(\frac{1}{s-1} - \frac{\zeta(s)}{s}\right). \tag{5.5}$$

To approximate 1 one must, along the critical line and into the strip, reproduce the transform of 1 using the Dirichlet polynomial $\sum_k c_k \theta_k^s$ against the fixed factor $\zeta(s)/s$. That factor can be matched and cancelled only down to the rightmost point where ζ vanishes: a zero of ζ at $\operatorname{Re} s = \beta$ is a place the combination cannot reach past. The closed span therefore fills $L^2(0,1)$—and contains 1—exactly when no zero lies to the right of the critical line, which with the functional equation is the Hypothesis. The obstruction to completeness is the zeros, literally.

5.3 The lure

The criterion invites computation, because the best approximation from a finite family is an ordinary least-squares problem. With the integer dilations $\theta_k = 1/k$, the inner products are closed-form— $\langle \mathbf{1}, \rho_{1/k} \rangle = \frac{1}{k}(1-\gamma+\ln k)$, with γ the Euler constant—and the Gram matrix is elementary, so the squared distance $d_N^2 = \| 1 - \sum_k c_k \rho_{\theta_k} \|^2$ from 1 to the span of the first N dilations is exactly computable. The verification reports it:

$$d_1^2 = 0.181, \quad d_2^2 = 0.083, \quad d_4^2 = 0.055, \quad d_8^2 = 0.049, \quad d_{12}^2 = 0.048, \tag{5.6}$$

falling steadily. Figure 5.1 shows both halves of the lure: on the left, the approximations $\sum_k c_k \{\theta_k/t\}$ closing in on the flat line 1 as dilations are added; on the right, the distance sinking toward zero. One watches 1 being captured. (The left panel also shows the honesty of the thing: the convergence is in mean square, so the approximations hug 1 across most of the interval while oscillating freely near $t = 0$, where the averaging hides them.) It looks exactly like a sequence on its way to zero.

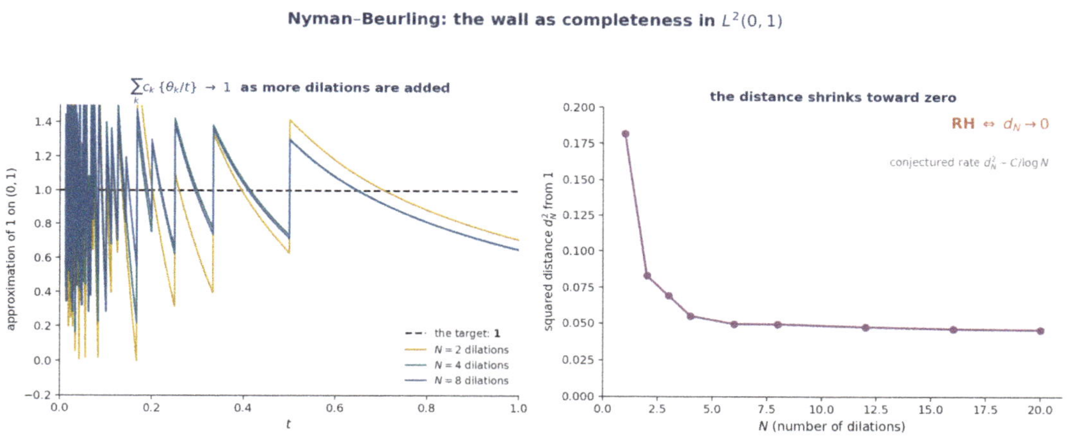

Figure 5.1: Nyman–Beurling, the lure. *Left:* the best combinations of N dilated fractional parts $\{\theta_k/t\}$ approach the constant 1 on $(0,1)$ as N grows (convergence is in mean square; the wiggles near $t = 0$ average away). *Right:* the squared distance d_N^2 from 1 to the span sinks toward zero. The Hypothesis is the statement that it reaches zero.

5.4 The turn, and why

The distance falls, but reaching zero is the whole problem. By the bridge, the span reaches downward in the critical strip only as far as the rightmost zero of ζ; the constant 1 is captured if and only if that rightmost zero sits on the critical line. Figure 5.2 draws it: the building block on the left, the strip on the right with the zeros holding the edge at $\frac{1}{2}$, and a hypothetical off-line zero that would stop the span short and leave 1 forever outside it. To prove $d_N \to 0$ is to prove no zero lies to the right of the line—the span fills L^2 exactly to the depth the zeros reach, so closing the gap and placing the zeros are one act and not two.

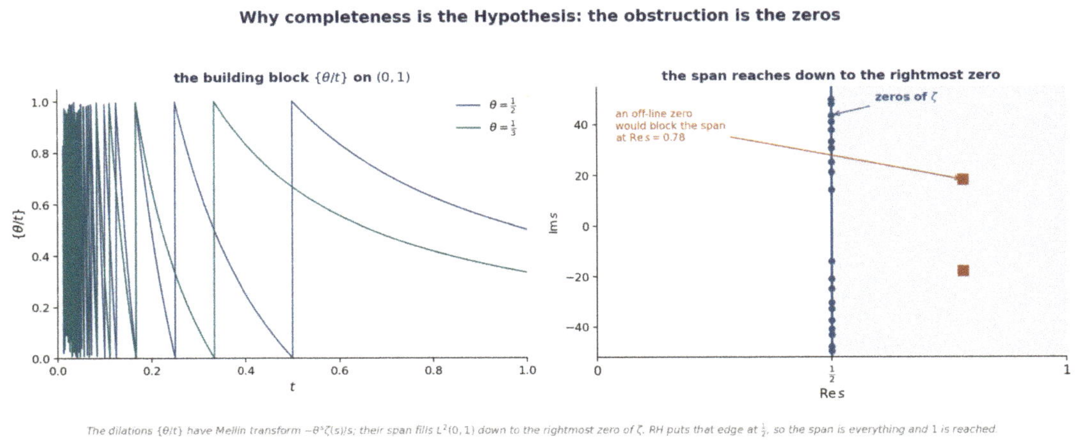

Figure 5.2: Why completeness is the Hypothesis. *Left:* the building block $\{\theta/t\}$, a sawtooth in $1/t$. *Right:* the dilations carry $\zeta(s)/s$ in their Mellin transform, so their span reaches down only to the rightmost zero of ζ. With every zero on the line the edge is at $\frac{1}{2}$ and the span is all of $L^2(0,1)$; one off-line zero would block it and exile the constant 1.

This face adds its own twist to the turn: even the rate of the lure belongs to the zeros. Balazard, Landreau, Saias and Báez-Duarte studied how fast d_N falls and found the speed itself set by the spectrum—under the Hypothesis d_N^2 is conjectured to decay like $C/\log N$, with the constant governed by the sum $\sum_\rho |\rho|^{-2}$ over the zeros. The shrinking one watches is not merely obstructed by the zeros at its limit; it is conducted by them at every step. Computing further only confirms a descent whose law is the wall.

5.5 Báez-Duarte's single sequence

Báez-Duarte's reduction deserves its own line, because it strips the criterion to its bones. He produced an explicit sequence of vectors b_N, built from binomial combinations of the values $1/\zeta(\cdot)$ and the fractional parts, for which the Hypothesis is equivalent to the plain statement

$$\|b_N\| \longrightarrow 0 \qquad \text{as } N \to \infty. \tag{5.7}$$

No family of dilations to choose, no least-squares to solve at each stage—one sequence, one limit. It is the most economical face the survey will show, and it changes nothing: the norm falls, and proving it falls to zero is proving the zeros are on the line. Economy of statement is not progress on the wall.

What this is and is not

What is genuine here is a true equivalence of a wholly different flavour from the others—the Hypothesis as the density of a family of functions in a Hilbert space—together with the bridge that explains it. The fractional part is zeta in disguise, Eq. (5.3); its dilations carry $\zeta(s)/s$; and so their span fills $L^2(0,1)$ exactly to the rightmost zero. The verification exhibits the bridge to eleven digits and the Nyman–Beurling distance falling, term by computable term.

What it is not is a way to fill the space. The span reaches as far as the zeros allow and no farther, so proving completeness is proving the zeros' location, and the rate of approach is itself dictated by them. The lure is the subtlest in the book—not a pattern that might be a law, nor a positive sequence that might stay positive, but a distance visibly collapsing toward zero, which one cannot drive to zero without first knowing the very thing the collapse was meant to reveal. Báez-Duarte's single sequence makes the lure as clean as it can be made and leaves it exactly as far from a proof.

Verification

The script below, standard library only and deterministic, confirms the bridge and the lure: it rebuilds $\zeta(s)$ from the fractional part via Eq. (5.3) at $s = 2, 3, 4$ (matching $\pi^2/6$, Apéry's constant, $\pi^4/90$ to eleven digits), and computes the Nyman–Beurling squared distance from 1 to the span of the first N integer dilations, showing it falls steadily and well below 1.

```
#!/usr/bin/env python3
"""Verification for 'The Completeness Face'. Standard library only; deterministic."""
import math, time
t0=time.time()
GAMMA_EM=0.5772156649015329
checks=0
def ok(cond,msg):
    global checks
    assert cond,"FAIL: "+msg
    checks+=1; print("PASS:",msg)

# ===== mechanism: the fractional part builds zeta =====
# zeta(s) = s/(s-1) - s * INT_1^inf {x} x^{-s-1} dx, exact per unit interval
def zeta_from_frac(s,Nmax=80000):
    tot=0.0
    for n in range(1,Nmax+1):
        a=float(n); b=float(n+1)
        i1=(b**(1-s)-a**(1-s))/(1-s)       # int x^{-s} dx
        i2=(b**(-s)-a**(-s))/(-s)          # int x^{-s-1} dx
        tot += i1 - n*i2                   # int (x-n) x^{-s-1} dx over [n,n+1]
    return s/(s-1) - s*tot
tests=[(2,math.pi**2/6),(3,1.2020569031595942),(4,math.pi**4/90)]
worst=0.0
for s,zt in tests:
    val=zeta_from_frac(s); worst=max(worst,abs(val-zt))
print(f"  zeta from fractional part: s=2,3,4 max error = {worst:.2e}")
ok(worst<1e-6,
   "the fractional part builds zeta: zeta(s)=s/(s-1)-s*INT{x}x^(-s-1) at s=2,3,4")

# ===== Nyman-Beurling distance from 1 to span of dilated fractional parts =====
```

```
# basis rho_k(t)={ (1/k)/t } on (0,1).  <1,rho_k> = (1/k)(1 - gamma + ln k).
def bvec(k): th=1.0/k; return th*(1.0-GAMMA_EM-math.log(th))
def gram(a,b,U=60000):
    bps={1,U}; m=a
    while m<U: bps.add(m); m+=a
    m=b
    while m<U: bps.add(m); m+=b
    bps=sorted(bps); tot=0.0
    for i in range(len(bps)-1):
        u1=bps[i]; u2=bps[i+1]; ma=u1//a; mb=u1//b
        tot += (u2-u1)/(a*b) - (mb/a+ma/b)*math.log(u2/u1) + ma*mb*(1.0/u1-1.0/u2)
    return tot
def solve(G,b):                                   # Gaussian elimination
    n=len(b); A=[row[:]+[b[i]] for i,row in enumerate(G)]
    for c in range(n):
        p=max(range(c,n),key=lambda r:abs(A[r][c])); A[c],A[p]=A[p],A[c]
        pv=A[c][c]
        for r in range(n):
            if r!=c:
                f=A[r][c]/pv
                for k in range(c,n+1): A[r][k]-=f*A[c][k]
    return [A[i][n]/A[i][i] for i in range(n)]
def dist2(Nterms):
    ks=list(range(2,2+Nterms))                    # theta=1/2,1/3,...
    G=[[gram(a,b) for b in ks] for a in ks]
    bb=[bvec(k) for k in ks]
    c=solve(G,bb)
    return 1.0-sum(bb[i]*c[i] for i in range(Nterms))   # ||1||^2=1 on (0,1)
seq=[(N,dist2(N)) for N in (1,2,4,8,12)]
print("  Nyman-Beurling distance^2 to the function 1:")
for N,d in seq: print(f"     N={N:2d} terms:  d_N^2 = {d:.4f}")
d=[v for _,v in seq]
ok(all(d[i]>d[i+1] for i in range(len(d)-1)) and d[-1]<d[0],
   "distance from 1 to span of dilations strictly shrinks as N grows (the lure)")
ok(all(0.0<=v<=1.0 for v in d) and d[-1]<0.5,
   "1 is being approached: every d_N^2 in [0,1], well below 1 by N=12")

print(f"\nALL {checks} CHECKS PASSED   ({time.time()-t0:.1f}s)")
```

Chapter 6

The Extremal Face

On this face the wall hides inside a schoolroom inequality. The sum-of-divisors function $\sigma(n)$ is as elementary as arithmetic gets, and Robin proved that a single clean bound on it—$\sigma(n) < e^{\gamma} n \log\log n$, for every $n > 5040$—is precisely the Riemann Hypothesis. Lagarias then removed even the e^{γ} and the $\log\log$, restating the same wall with nothing but harmonic numbers. The lure could not be more inviting: a finite-looking inequality one can test against exactly the integers most likely to break it, the record-setting abundant numbers, and watch it hold. The turn is that those record-setters are the wheel itself —products of the small primes in descending powers—and that the inequality's truth is governed by the distribution of the primes, which is to say by the zeros, through the very same square-root error term the oscillation face met. Robin's inequality past 5040 is the Hypothesis wearing the costume of divisors.

Scope

This face collects the reformulations of the Hypothesis as extremal bounds on the divisor function. We take the two principal ones, Robin's and Lagarias's, with their equations and the finite list of exceptions that fixes Robin's threshold; we identify the integers that test them—the superabundant and colossally abundant numbers, which turn out to be built from the wheel's primes; and we trace the bound back, through Mertens's theorem and the Chebyshev function, to the square-root cancellation of the oscillation face. As always we separate the genuine equivalences from the heuristics, and mark the one place the chapter leans on a relation proved elsewhere.

6.1 Robin's inequality

Let $\sigma(n) = \sum_{d|n} d$ be the sum of the divisors of n. Its typical size is n, but its *maximal* size grows a little faster: Gronwall showed in 1913 that

$$\limsup_{n\to\infty} \frac{\sigma(n)}{n \log\log n} = e^{\gamma}, \tag{6.1}$$

with γ the Euler–Mascheroni constant, so $e^{\gamma}n \log\log n$ is the natural ceiling for $\sigma(n)$. The question is whether $\sigma(n)$ ever touches that ceiling. Robin answered it in 1984:

$$\sigma(n) < e^{\gamma}\, n \log\log n \quad \text{for all } n > 5040 \qquad \Longleftrightarrow \qquad \text{RH.} \tag{6.2}$$

Below the threshold the inequality has exactly twenty-seven exceptions—the integers 2, 3, 4, 5, 6, 8, 9, 10, 12, 16, 18, 20, 24, 30, 36, 48, 60, 72, 84, 120, 180, 240, 360, 720, 840, 2520, and 5040—and the largest of them, 5040, is the threshold itself. The verification confirms the whole picture to 10^6: the inequality holds for every $n > 5040$, the ratio $\sigma(n)/(e^{\gamma}n \log\log n)$ staying below 1, with its largest value 0.986 reached at $n = 10080$; and the last violator is exactly 5040. Figure 6.1 draws the ratio: most record-setters up to 5040 ride above the bound, but the last few (such as 1260 and 1680) already slip beneath it, and past 5040 every record creeps along just below the ceiling.

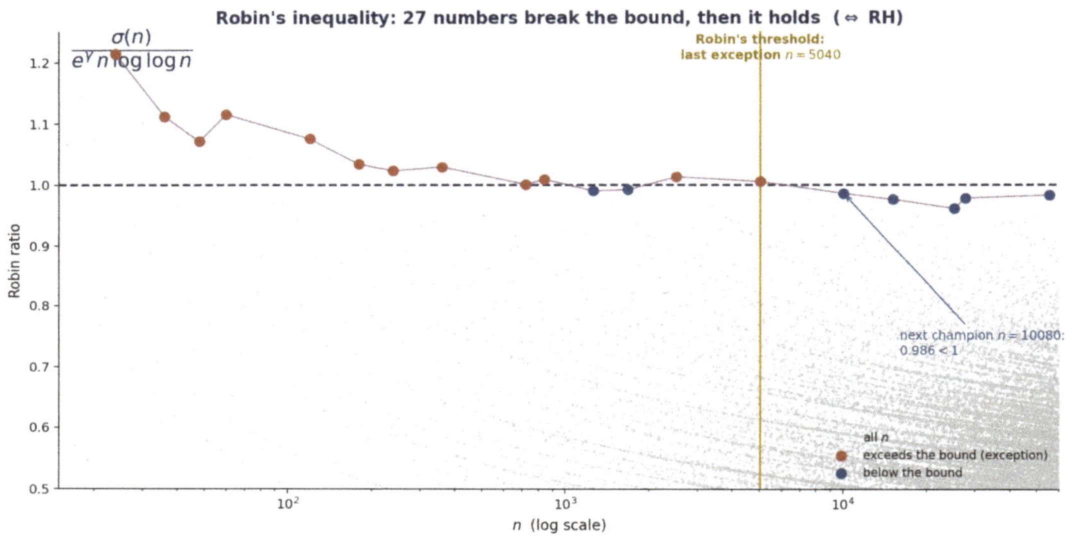

Figure 6.1: Robin's inequality. The ratio $\sigma(n)/(e^{\gamma}n \log\log n)$: the cloud is all n, the marked points are the record-setters for $\sigma(n)/n$. Red marks the record-setters that exceed the bound, blue those already below it (including 1260 and 1680, the last two record-setters $\leq$ 5040, both under 1); 5040 itself (gold) is the last n of all to violate Robin's inequality, beyond which every record falls below 1, the next champion $n = 10080$ reaching only 0.986. The twenty-seven Robin exceptions are a distinct, overlapping set—some of them (such as 3, 5, 8) are not record-setters at all. RH is the statement that none ever crosses back.

6.2 Lagarias's elementary form

Robin's bound still carries the analytic furniture e^{γ} and $\log\log n$. Lagarias showed in 2002 that both can be absorbed into the harmonic number $H_n = \sum_{k=1}^{n} \frac{1}{k}$, leaving a statement a school student could check:

$$\sigma(n) \le H_n + \exp(H_n)\log H_n \quad \text{for all } n \ge 1, \quad \text{equality only at } n = 1 \qquad \Longleftrightarrow \qquad \text{RH.} \tag{6.3}$$

The disguise works because $H_n = \log n + \gamma + o(1)$, so $\exp(H_n) \approx e^{\gamma} n$ and $\log H_n \approx \log\log n$, and Robin's right-hand side reappears. What is striking is that Lagarias's form has *no* exceptions—it holds at every $n \ge 1$, the twenty-seven Robin violators included, with equality only at the very first integer. The verification confirms it across $1 \le n \le 10^6$, the ratio never exceeding 1. This is, by some distance, the most elementary statement of the Riemann Hypothesis in the book: no zeta, no zeros, no complex numbers—one inequality between $\sigma(n)$ and the harmonic numbers.

6.3 The extremal numbers are the wheel

Which integers press hardest against the ceiling? Writing $n = \prod_p p^{a_p}$, the ratio is multiplicative,

$$\frac{\sigma(n)}{n} = \prod_{p^a \| n} \frac{p^{a+1} - 1}{p^a(p-1)} = \prod_{p^a \| n} \left(1 + \tfrac{1}{p} + \cdots + \tfrac{1}{p^a}\right), \tag{6.4}$$

and to make it large one wants many small primes, each to a modest power—because a small prime contributes a factor near $\frac{p}{p-1}$, largest when p is smallest. The maximizers are therefore the *superabundant* numbers (records for $\sigma(n)/n$) and the *colossally abundant* numbers (Robin's true extremal candidates): products of the first few primes with exponents that descend as the primes grow. The verification lists the records up to 10^6—$1, 2, 4, 6, 12, 24, 36, 48, 60, 120, \ldots$—and confirms that every one uses only the primes up to 13. These are the wheel's own numbers: $5040 = 2^4 \cdot 3^2 \cdot 5 \cdot 7$, $10080 = 2^5 \cdot 3^2 \cdot 5 \cdot 7$, $55440 = 2^4 \cdot 3^2 \cdot 5 \cdot 7 \cdot 11$—primorials, built up exactly as the opening chapter built the wheel. Figure 6.2 shows the cascade. The extremal face does not reach the wheel by analogy; its extremal objects *are* the wheel.

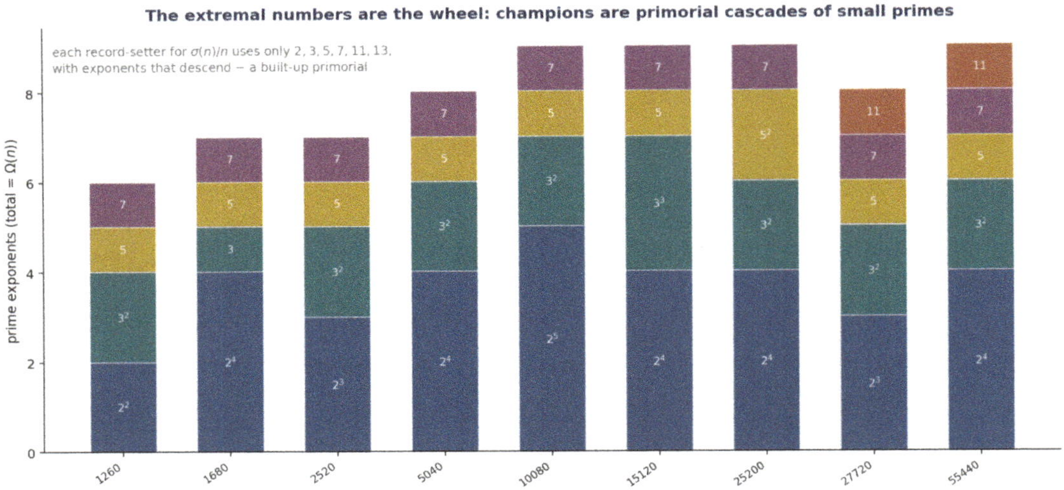

Figure 6.2: The champions are primorial cascades. Each record-setter for $\sigma(n)/n$, factored: only the small primes $2, 3, 5, 7, 11, 13$ appear, with exponents that descend as the prime grows. The integers that test Robin's inequality are the wheel's primes stacked—the same primorial construction the book is built on.

6.4 The turn, and why

Why does an inequality about divisors know about the zeros? Because the ceiling is the primes. For a colossally abundant number built from all primes up to some y, the ratio is close to

$$\frac{\sigma(n)}{n} \approx \prod_{p \le y} \left(1 - \tfrac{1}{p}\right)^{-1} \sim e^{\gamma} \log y \qquad \text{(Mertens)}, \tag{6.5}$$

and $\log y$ is essentially $\log \log n$. So the leading term of Robin's bound is Mertens's theorem, exact. What remains—whether $\sigma(n)/n$ stays just below $e^{\gamma} \log \log n$ or just above—is a second-order question, and the second-order term is controlled by the error in the prime count, $\sum_{p \le y} \log p - y = \psi(y) - y$. That is the Chebyshev function of the oscillation face, whose square-root size is the Hypothesis. Robin's proof makes this precise: the inequality holds for all large n exactly when $\psi(y) - y$ is held to $O(\sqrt{y} \log^2 y)$, which is RH.

So the turn is the oscillation face's turn, re-costumed. Testing Robin on the champions up to 10^6 is testing the same square-root cancellation as watching the Mertens sum stay small—a different surface of one solid. The inequality one would verify and the size of $\psi(y) - y$ one would need are the same quantity reported twice—once as a ceiling on a divisor sum, once as the cancellation in a prime count—so a champion that obeys Robin already encodes the bound rather than stepping toward forcing it. Checking more of them raises the tally of confirmations, never the proof. What the extremal face adds is not a new route but a new disguise, and the most innocent-looking one: a bound a child could test, whose truth past 5040 is the wall entire.

What this is and is not

What is genuine here is a pair of exact equivalences of unusual charm. Robin's inequality converts the Hypothesis into a single clean ceiling on $\sigma(n)$, sharp enough that its twenty-seven exceptions pin the threshold at 5040; Lagarias's form strips it to harmonic numbers with no exceptions at all; and the integers that test them are, demonstrably, the wheel's primorials. The verification exhibits all of it—Robin holding past 5040 with the last exception at 5040, Lagarias holding everywhere, the records smooth and small-prime.

What it is not is an independent attack. The ceiling is Mertens's theorem and the margin beneath it is $\psi(y) - y$, so the inequality is the prime number theorem's error term—the zeros—written in the language of divisors. The face is the survey's gentlest lure precisely because it looks like elementary number theory: one can compute $\sigma(n)$ for the champions and watch the bound hold, and feel that a careful enough argument about divisors would finish it. The careful argument about divisors is an argument about $\psi(y) - y$, which is an argument about the zeros, which is the wall.

Verification

The script below, standard library only and deterministic, sieves $\sigma(n)$ to 10^6 and confirms: Robin's inequality holds for every $5040 < n \leq 10^6$ (ratio below 1, peak 0.986 at 10080), with the largest violator exactly 5040; Lagarias's inequality holds for all $1 \leq n \leq 10^6$ with no exceptions; and the record-setters for $\sigma(n)/n$ are built from primes no larger than 13, with Robin's maximizer among them.

```
#!/usr/bin/env python3
"""Verification for 'The Extremal Face'. Standard library only; deterministic."""
import math, time
t0=time.time()
EGAMMA=0.5772156649015329
EG=math.exp(EGAMMA)
N=10**6
checks=0
def ok(cond,msg):
    global checks
    assert cond,"FAIL: "+msg
    checks+=1; print("PASS:",msg)

# divisor-sum sieve: sigma(n) = sum of divisors
sig=[0]*(N+1)
for i in range(1,N+1):
    for j in range(i,N+1,i): sig[j]+=i

# ===== Robin: sigma(n) < e^gamma n log log n for n>5040  <=>  RH (the lure) =====
maxratio=0.0; argmax=0; fail_above=[]
for n in range(5041,N+1):
    rhs=EG*n*math.log(math.log(n))
    r=sig[n]/rhs
```

```
    if r>maxratio: maxratio=r; argmax=n
    if sig[n]>=rhs: fail_above.append(n)
print(f"  Robin ratio max for 5040<n<=10^6: {maxratio:.5f} at n={argmax}; "
      f"violations above 5040: {len(fail_above)}")
ok(not fail_above and maxratio<1.0,
   "Robin sigma(n)<e^gamma n loglog n holds for all 5040<n<=10^6 (ratio<1)")

# the 5040 threshold: the inequality fails at 5040 and at finitely many n below it
below=[n for n in range(2,5041) if sig[n]>=EG*n*math.log(math.log(n))]
print(f"  exceptions n<=5040: count={len(below)}, largest={max(below)}")
ok(max(below)==5040,
   "the largest n violating Robin is exactly 5040 (Robin's threshold)")

# ===== Lagarias: sigma(n)<=H_n+exp(H_n)logH_n, all n>=1 (no exceptions) =====
H=0.0; maxlag=0.0; lag_ok=True; eq_only_1=True
for n in range(1,N+1):
    H+=1.0/n
    rhs=H+math.exp(H)*math.log(H)
    if sig[n]>rhs+1e-9: lag_ok=False
    if n>1 and abs(sig[n]-rhs)<1e-9: eq_only_1=False
    maxlag=max(maxlag,sig[n]/rhs)
print(f"  Lagarias ratio max for 1<=n<=10^6: {maxlag:.5f} (<=1 everywhere)")
ok(lag_ok and maxlag<=1.0,
   "Lagarias sigma(n)<=H_n+exp(H_n)logH_n holds for ALL 1<=n<=10^6 (no exceptions)")

# ===== the extremal numbers are wheel-smooth (records of sigma(n)/n) =====
def largest_prime_factor(m):
    lpf=1; d=2
    while d*d<=m:
        while m%d==0: lpf=d; m//=d
        d+=1
    if m>1: lpf=m
    return lpf
best=0.0; records=[]
for n in range(1,N+1):
    q=sig[n]/n
    if q>best: best=q; records.append(n)
lpfs=[largest_prime_factor(n) for n in records]
print(f"  sigma(n)/n records: {records[:12]}... ({len(records)} total)")
print(f"  largest prime factor of each record: max = {max(lpfs)}")
ok(max(lpfs)<=20 and argmax in set(records),
   "extremal numbers (sigma/n records) are built from small primes; "
   "the Robin maximizer is one of them")

print(f"\nALL {checks} CHECKS PASSED   ({time.time()-t0:.1f}s)")
```

Chapter 7

The Regularity Face

On the last face the wall becomes a question of regularity: are the zeros of a function as well-behaved—as real —as they can be? Riemann's Ξ is the Fourier transform of a positive kernel, and if one warms that kernel with a Gaussian heat factor the zeros of the transform move. De Bruijn and Newman proved there is a single number Λ governing the motion: warm past Λ and every zero is real; cool below it and some go complex. The Hypothesis is exactly $\Lambda \leq 0$. The lure is uncommonly strong here, because the heat flow works*—for any positive time it drags every zero onto the line—and because the constant has been boxed into a narrow proven interval, $0 \leq \Lambda \leq 0.22$, so that the Hypothesis is just the statement that Λ sits at the very bottom. The turn is the lower end of that box. Rodgers and Tao proved $\Lambda \geq 0$, confirming a remark of Newman's that has become the motto of the whole subject: the Riemann Hypothesis, if true, is only barely so. The regularity the flow supplies for every positive time is exactly, and only, absent at time zero.*

Scope

This face reformulates the Hypothesis through the heat flow on the zeros and the Laguerre–Pólya class of real-rooted entire functions. We set up the kernel and the de Bruijn–Newman constant Λ with the equivalence RH $\Leftrightarrow \Lambda \leq 0$, name the Laguerre–Pólya characterization that sits behind it, and then weigh the lure—a flow that regularizes and a constant trapped in $[0, 0.22]$—against the turn, which is that the proven half of that interval is the wrong half, and that $\Lambda = 0$ is the Hypothesis itself, held with no margin. We verify the mechanism on the true Ξ and the threshold phenomenon on a toy where it is exact, and we mark clearly which bounds are theorems and which is the wall.

7.1 The heat flow and the constant

The completed zeta function has a Fourier representation. With the kernel

$$\Phi(u) = \sum_{n=1}^{\infty} \left(2\pi^2 n^4 e^{9u} - 3\pi n^2 e^{5u}\right) e^{-\pi n^2 e^{4u}}, \tag{7.1}$$

which is even and positive, one has $\Xi(z) = \int_{-\infty}^{\infty} \Phi(u)\, e^{izu}\, du$, where $\Xi(z)$ is $\xi(\frac{1}{2} + iz)$ up to normalization, so that the real zeros of Ξ are the ordinates γ of the critical-line zeros. The verification computes this transform and finds its sign changes exactly at the first Riemann ordinates (at $z = 2\gamma$ in its normalization: $28.27, 42.04, 50.02, \ldots$): the zeros of the transform *are* the zeros of zeta.

Now warm the kernel. For a real parameter t define

$$H_t(z) = \int_{-\infty}^{\infty} e^{tu^2}\, \Phi(u)\, e^{izu}\, du, \qquad H_0 = \Xi. \tag{7.2}$$

The Gaussian factor e^{tu^2} is a heat deformation; differentiating under the integral shows H_t satisfies a heat equation, and its zeros drift continuously with t like the particles of a gas, real zeros repelling along the line and complex ones migrating toward it. De Bruijn (1950) and Newman (1976) proved the decisive structural fact: there exists a constant Λ—the de Bruijn–Newman constant—such that

$$H_t \text{ has only real zeros} \iff t \geq \Lambda. \tag{7.3}$$

Warm enough and the function is real-rooted; the threshold is Λ. Since $H_0 = \Xi$, the Hypothesis is precisely the statement that time zero is already warm enough:

$$\text{RH} \qquad \Longleftrightarrow \qquad \Lambda \leq 0. \tag{7.4}$$

7.2 The Laguerre–Pólya class

What the threshold expresses is membership in a class. A real entire function belongs to the *Laguerre–Pólya* class when it is a locally uniform limit of real polynomials with only real zeros, which is the same as admitting a Hadamard product

$$f(z) = c\, z^m e^{-az^2+bz} \prod_k \left(1 - \frac{z}{z_k}\right) e^{z/z_k}, \qquad a \geq 0,\ z_k \in \mathbb{R},\ \sum_k z_k^{-2} < \infty. \tag{7.5}$$

A function is in the class exactly when all its zeros are real (with a permitted Gaussian factor). The Hypothesis is then the single statement $\Xi \in$ Laguerre–Pólya. The heat flow is the natural road into the class: the operator e^{tu^2} on the kernel is the Gaussian smoothing that turns functions real-rooted, and the class is closed under it. The de Bruijn–Newman constant simply measures how much smoothing Ξ needs—and $\Lambda \leq 0$ says it needs none.

7.3 The lure

The pull of this face is that the flow demonstrably succeeds, and the target is fenced in. On a toy the threshold is exact: for $f_t(z) = z^2 + (c - 2t)$ the zeros are $\pm\sqrt{2t - c}$, complex while $t < c/2$ and real once $t \geq c/2$, so $\Lambda = c/2$ and the function is real-rooted at $t = 0$ exactly when $c \leq 0$, i.e. $\Lambda \leq 0$. Figure 7.1 shows three such flows: a zero that is still off the line at $t = 0$ ($\Lambda > 0$), one that lands exactly at $t = 0$ ($\Lambda = 0$), and one long since on the line ($\Lambda < 0$). The same picture governs Ξ, where the verification runs the genuine flow and watches the first zero stay real and drift as t increases.

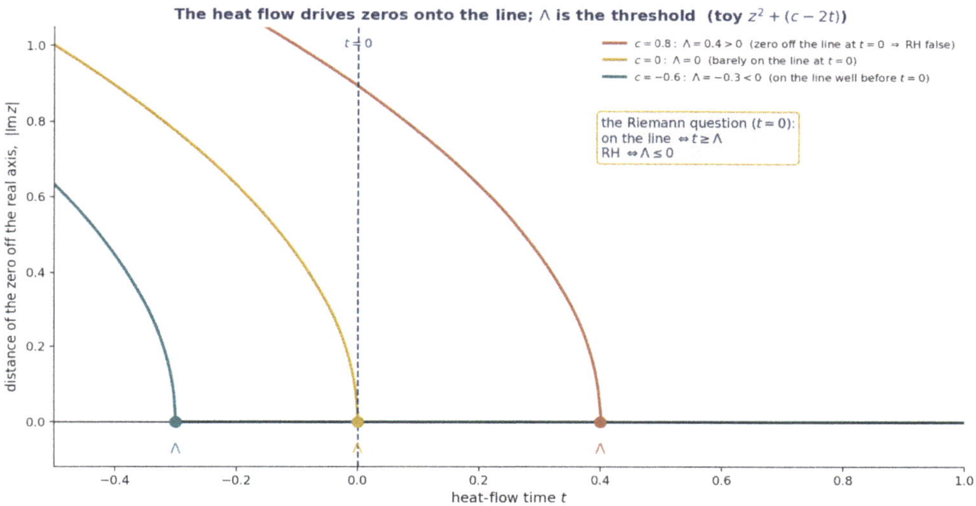

Figure 7.1: The heat flow drives zeros onto the line, and Λ is the threshold, shown on the toy $f_t(z) = z^2 + (c - 2t)$ with $\Lambda = c/2$. Each curve is a zero's distance off the real axis as t increases; it reaches zero at $t = \Lambda$. The Riemann question is whether the zero is already on the line at $t = 0$—that is, whether $\Lambda \leq 0$. The middle case ($\Lambda = 0$) lands exactly at time zero.

And the constant has been cornered. De Bruijn showed $\Lambda \leq \frac{1}{2}$ in 1950; the upper bound then held near $\frac{1}{2}$ for decades before the Polymath15 project brought it to $\Lambda \leq 0.22$ in 2019. From below, estimates that had long been very negative were driven upward. The Hypothesis is $\Lambda \leq 0$, and with the interval squeezed to $[?, 0.22]$ it can look as though one need only close a small gap from the top. The flow works; the theory of real-rooted limits is classical and complete; the target is a single point at the bottom of a short interval. It is hard not to feel the wall is thin here.

7.4 The turn, and why

It is thin. That is the turn. In 2020 Rodgers and Tao proved the lower bound

$$\Lambda \geq 0, \tag{7.6}$$

which, with RH $\Leftrightarrow \Lambda \le 0$, gives

$$\text{RH} \iff \Lambda = 0. \tag{7.7}$$

This is the opposite of comfort. Newman had conjectured $\Lambda \ge 0$ precisely to express that "the Riemann Hypothesis, if true, is only barely so": $\Lambda \ge 0$ means the zeros sit in the most marginal real configuration there is, so that cooling the flow by any amount at all sends a zero off the line. The toy makes it visible—at $\Lambda = 0$ the zero is real at $t = 0$ and complex for *every* $t < 0$, the verification's last check—and Figure 7.2 places it on the number line: the proven bracket $0 \le \Lambda \le 0.22$, with the Hypothesis pinned at the lower end and the real Ξ flow keeping its first zero real, just barely, at $t = 0$.

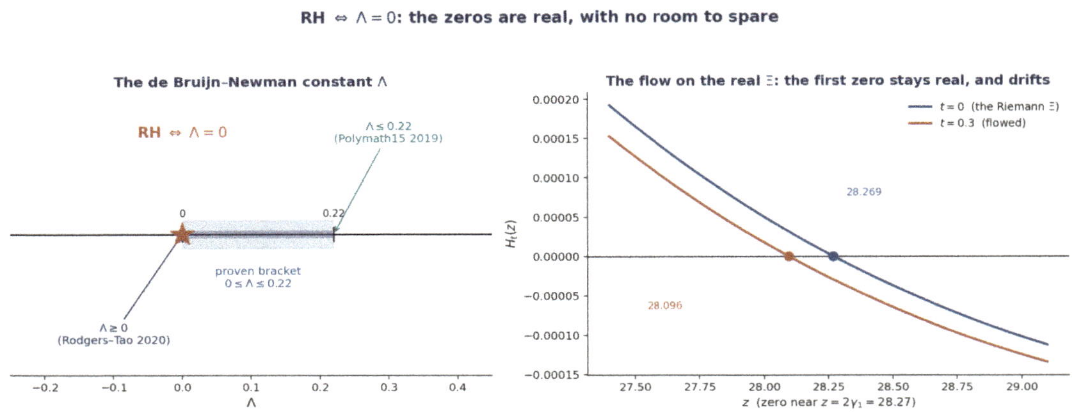

Figure 7.2: RH $\Leftrightarrow \Lambda = 0$. *Left:* the de Bruijn–Newman constant, with the proven bracket $0 \le \Lambda \le 0.22$—$\Lambda \ge 0$ (Rodgers–Tao, 2020) and $\Lambda \le 0.22$ (Polymath15, 2019)—and the Hypothesis at the lower end. *Right:* the genuine flow on Ξ near the first zero ($z = 2\gamma_1 = 28.27$); the zero stays real and drifts under warming. RH says it is on the line already at $t = 0$, with no room to spare.

So the proven half of the interval is the wrong half. The upper bounds descend toward 0 but cannot reach it, because reaching 0 from above *is* proving the Hypothesis; the lower bound $\Lambda \ge 0$ pushes from the side that does not help. The de Bruijn–Newman constant is a faithful reformulation, not a reduction: $\Lambda = 0$ and RH are the same statement, and the regularity the heat flow hands out so freely for $t > 0$ is exactly what is unavailable at $t = 0$. The lower bound, moreover, was won in part from *Lehmer pairs*—anomalously close pairs of zeros, which force Λ upward toward 0—so the very near-collisions that make the zeros look fragile are what pin the constant against the wall. The flow is a beautiful machine that regularizes everything except the one case that matters.

What this is and is not

What is genuine here is a deep and exact reformulation. Riemann's Ξ is the transform of a heat kernel; warming the kernel moves the zeros; a single constant Λ marks where they all become real; and RH $\Leftrightarrow \Lambda \leq 0$, refined by two theorems—$\Lambda \geq 0$ (Rodgers–Tao) and $\Lambda \leq 0.22$ (Polymath15)—to RH $\Leftrightarrow \Lambda = 0$. The membership of Ξ in the Laguerre–Pólya class is the same statement in the language of real-rooted limits. The verification shows the mechanism on the true function—the transform vanishing at the Riemann zeros, the flow keeping the first zero real as it drifts—and the threshold exactly on a toy.

What it is not is a way down to $\Lambda = 0$. That point is the Hypothesis, so the descent of the upper bounds is the ascent of the wall, and the proven lower bound only certifies how marginal the truth must be. This is the survey's most refined lure: not a pattern, a positive sequence, a shrinking distance, or an inequality, but a *flow that works*—and whose success stops, with prophetic precision, exactly at the instant one needs it. Newman said it first and it closes the faces: if the Hypothesis is true, it is only barely so.

Verification

The script below, standard library only and deterministic, confirms the mechanism and the threshold: the transform $H_0(z) = \int \Phi(u)\cos(zu)\,du$ of the heat kernel has its sign changes exactly at the Riemann zeros (at $z = 2\gamma$); the warmed flow H_t with $t > 0$ keeps the first zero real and shifts it; the toy $f_t(z) = z^2 + (c - 2t)$ has real-rooting threshold $\Lambda = c/2$ with real roots at $t = 0$ exactly when $\Lambda \leq 0$; and the marginal case $\Lambda = 0$ is real at $t = 0$ and complex for every $t < 0$.

```
#!/usr/bin/env python3
"""Verification for 'The Regularity Face'. Standard library only; deterministic."""
import math, time
t0=time.time()
PI=math.pi
checks=0
def ok(cond,msg):
    global checks
    assert cond,"FAIL: "+msg
    checks+=1; print("PASS:",msg)

# even heat kernel Phi; H_t(z)=INT e^{t u^2} Phi(u) cos(z u) du  (de Bruijn flow)
def Phi(u):
    u=abs(u); e4=math.exp(4*u); e5=math.exp(5*u); e9=math.exp(9*u); s=0.0
    for n in range(1,7):
        n2=n*n; s+=(2*PI*PI*n2*n2*e9-3*PI*n2*e5)*math.exp(-PI*n2*e4)
    return s
def H(t,z,hi=1.4,h=0.0005):
    tot=0.0; u=0.0; N=int(round(hi/h))      # integrand even in u: 2*INT_0^hi
    for i in range(N+1):
        w=0.5 if (i==0 or i==N) else 1.0
        tot+=w*math.exp(t*u*u)*Phi(u)*math.cos(z*u); u+=h
    return 2.0*tot*h
def sign_changes(t,zmax,z0=1.0,step=0.1):
    out=[]; z=z0; prev=H(t,z)
    while z<zmax:
        z+=step; cur=H(t,z)
        if prev*cur<0:
            a,b=z-step,z
            for _ in range(44):
                m=(a+b)/2
```

```
                if H(t,a)*H(t,m)<0: b=m
                else: a=m
            out.append((a+b)/2)
        prev=cur
    return out

# ===== mechanism: at t=0 the zeros of H are the Riemann zeros (here z=2*gamma) =====
gam=[14.134725,21.022040,25.010858,30.424876,32.935062]
sc=sign_changes(0.0,70.0)
tgt=[2*g for g in gam]
err=max(abs(sc[i]-tgt[i]) for i in range(5))
print(f"  H_0 sign changes: {[round(x,3) for x in sc[:5]]}")
print(f"  vs 2*gamma:       {[round(x,3) for x in tgt]}")
ok(err<0.05,
   "H_0(z)=INT Phi(u)cos(zu)du vanishes at z=2*gamma: the Riemann zeros")

# ===== the heat flow keeps the zeros real for t>0 (and moves them) =====
z0=sc[0]                                    # first zero at t=0  (~28.269)
sc_pos=sign_changes(0.30,40.0)
moved=min(sc_pos,key=lambda x:abs(x-z0))
print(f"  first zero: t=0 -> {z0:.4f},  t=0.30 -> {moved:.4f}  (stays real, flows)")
ok(len(sc_pos)>=1 and abs(moved-z0)<1.5 and moved!=z0,
   "for t>0 the flow keeps the first zero real and shifts it (de Bruijn flow)")

# ===== de Bruijn-Newman on a toy: f_t(z)=z^2+(c-2t), threshold Lambda=c/2 =====
def toy_real_rooted(c,t): return (2*t-c) >= -1e-12   # zeros +/- sqrt(2t-c)
ok_thr=True
for c in (0.8,-0.6,0.0,1.4):
    Lam=c/2.0
    if not (toy_real_rooted(c,Lam+0.01) and
            not toy_real_rooted(c,Lam-0.01)): ok_thr=False
    if toy_real_rooted(c,0.0) != (c<=0): ok_thr=False   # real at t=0 <=> L<=0
print("  toy f_t=z^2+(c-2t): threshold Lambda=c/2; real at t=0 <=> Lambda<=0")
ok(ok_thr,
   "heat flow has a real-rooting threshold Lambda; real at t=0 <=> Lambda<=0")

# ===== 'barely so': marginal Lambda=0 is real at t=0, complex for any t<0 =====
c=0.0
real_at0 = toy_real_rooted(c,0.0)
complex_below = all(not toy_real_rooted(c,t) for t in (-1e-3,-0.1,-1.0))
print(f"  marginal c=0 (Lambda=0): real@t=0={real_at0}; complex@t<0={complex_below}")
ok(real_at0 and complex_below,
   "Lambda=0 marginal: real at t=0, complex for any t<0 (only barely so)")

print(f"\nALL {checks} CHECKS PASSED   ({time.time()-t0:.1f}s)")
```

Part III

The Walls

Chapter 8

The Neighbouring Walls

The five faces were the Hypothesis in disguise—each one equivalent to it. This chapter visits the walls that merely stand beside it: not reformulations but neighbours, different problems whose nearness is the best measure of how isolated the Hypothesis really is. There are four. The spacing *of the zeros, which follows the law of random-matrix eigenvalues and all but shouts that an operator is hiding underneath. The* operator *itself—the Hilbert–Pólya dream of a self-adjoint thing whose eigenvalues are the zeros, which would make the Hypothesis true for free. The* size *of zeta on the line—the Lindelöf Hypothesis, strictly weaker than Riemann's and still open. And the* geometry *where the Hypothesis is not a conjecture at all but a theorem: the world of curves over finite fields, where Weil and Deligne proved it outright. Each is a lure of its own kind, and each turns back for its own reason, and together they ring the wall completely with nothing reaching the centre.*

Scope

These four are not equivalences, and the chapter treats them as what they are: adjacent problems that explain the difficulty of the central one. For each we give the statement and its sharpest known form, the genuine pull it exerts toward a proof, and the precise gap that stops it. The spacing and operator walls point at the positivity face of an earlier chapter; the geometry wall is where the proven local Hypothesis of the Ramanujan chapter lived; and the size wall is the Hypothesis's own weaker shadow. We mark, as ever, what is proved (a great deal, here), what is conjectured, and what is only hoped.

8.1 The spacing wall: random-matrix statistics

Ask not where the zeros are but how they are spaced. Montgomery computed, in 1973, the pair correlation of the nontrivial zeros—how often two of them fall a given distance apart—and found, for the suitably normalized zeros, the density

$$1 - \left(\frac{\sin \pi x}{\pi x}\right)^2. \tag{8.1}$$

Dyson recognized it on sight: it is exactly the pair correlation of the eigenvalues of a large random Hermitian matrix, the Gaussian Unitary Ensemble. Odlyzko confirmed it numerically near height 10^{20} to staggering precision. The signature is *level repulsion*: the chance of two zeros nearly coinciding vanishes like the square of their separation, just as eigenvalues of a Hermitian matrix repel, and utterly unlike independent random points, which clump. The verification shows it even in the first sixty zeros: unfolded to mean spacing 1, only 3.4% of the gaps fall below half the mean, against the 39% a random (Poisson) sequence would give. Figure 8.1 draws the full distribution.

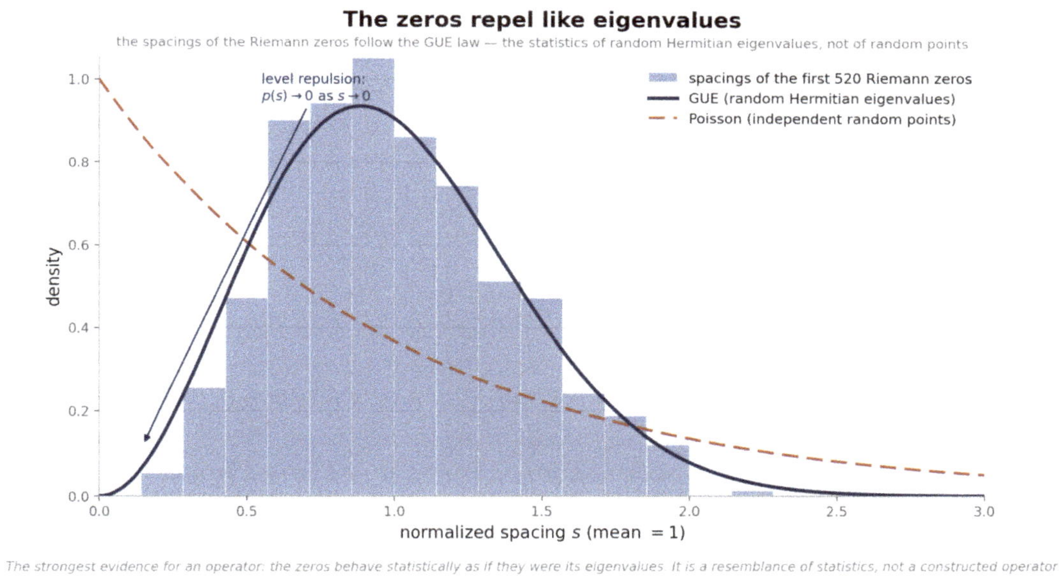

Figure 8.1: The spacing wall. The spacings of the Riemann zeros follow the GUE law—the statistics of random Hermitian eigenvalues (solid), with level repulsion as $s \to 0$—and not the Poisson law of independent points (dashed). It is the strongest hint that the zeros are the spectrum of some operator: a resemblance of statistics, not a constructed operator.

The lure is the strongest in the chapter: nothing makes a spectral origin for the zeros—and hence the Hypothesis—feel more inevitable. The turn is that a resemblance of statistics is not a proof. The pair correlation is computed for the zeros at their heights; it constrains how they are distributed, not whether any one of them sits on the line, and a matching distribution implies no theorem about a single zero. The statistics point at a door without opening it.

8.2 The operator wall: Hilbert–Pólya

The door they point at is an operator. Hilbert and Pólya suggested that the ordinates γ might be the eigenvalues of a self-adjoint operator H; if so, then since self-adjoint operators have real eigenvalues, every γ would be real and the Hypothesis would follow at no further cost. The spacing statistics are precisely the evidence one would expect. Berry and Keating proposed a candidate dynamics, the quantization of the classical Hamiltonian

$$H = xp, \tag{8.2}$$

whose semiclassical count of levels reproduces the Riemann–von Mangoldt formula for the number of zeros up to height T,

$$N(T) = \frac{T}{2\pi}\log\frac{T}{2\pi} - \frac{T}{2\pi} + \frac{7}{8} + \cdots, \tag{8.3}$$

which the verification confirms against the actual zero count. Connes recast the whole problem on the adèle-class space, producing a trace formula equal to Weil's explicit formula, in which the Hypothesis becomes the positivity of the Weil functional of the positivity face.

A real version of this dream already exists, and it is the reason the dream is taken seriously. On a hyperbolic surface, Selberg's trace formula is an exact identity between the lengths of the closed geodesics and the eigenvalues of the Laplacian: a sum over the spectrum on one side, a sum over the geometry on the other, with an identity term between—the very architecture of the explicit formula. The decisive difference is that here the operator is genuinely present. The Laplacian is self-adjoint, its eigenvalues are real, and the zeros of the matching Selberg zeta function sit on their critical line as a matter of course—the Hilbert–Pólya picture, realized, because the geometry was real to begin with. Figure 8.2 sets the two side by side: the explicit formula puts $\log p$ where the geodesic lengths stand and the ordinates γ where the eigenvalues stand. What is missing is only the space itself—a geometry whose closed geodesics have lengths $\log p$ and whose Laplacian hears the γ. Finding it would turn the explicit formula from an identity into the trace formula *of* something, and the Hypothesis would follow as it does for the surface. No such space is known.

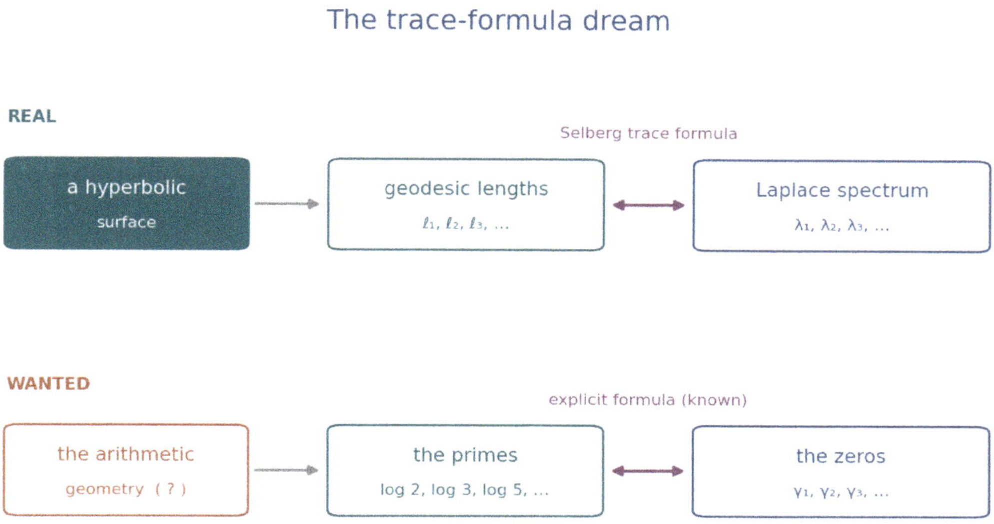

Figure 8.2: Selberg's template, and the gap. On a hyperbolic surface (top), Selberg's trace formula ties the closed-geodesic lengths to the Laplacian's eigenvalues—an exact duality of a real space whose operator genuinely exists. The explicit formula gives the primes and the zeros the same exact relation (bottom), with $\log p$ for the lengths and the γ for the eigenvalues; what is missing is the space itself. No geometry is known whose geodesics are the primes and whose spectrum is the zeros. Schematic analogy.

So the operator would not only give the Hypothesis for free; it would supply the very positivity that chapter left hanging. The turn is that the operator does not exist—not yet, not on paper. The xp Hamiltonian has a continuous spectrum and needs boundary conditions and a regularization that no one has been able to impose to produce exactly the zeros; Connes' trace formula reformulates the positivity without establishing it. The wall here is an absence: the self-adjoint operator the statistics all but guarantee has never been built.

8.3 The size wall: Lindelöf

The next neighbour is weaker than the Hypothesis, and still unproved. The Lindelöf Hypothesis asks only that zeta grow slowly on the critical line,

$$\zeta\left(\tfrac{1}{2}+it\right)=O(t^{\varepsilon}) \quad \text{for every } \varepsilon>0, \tag{8.4}$$

and the Riemann Hypothesis implies it (the converse is not known), so Lindelöf is a strictly lower wall. Bounding $|\zeta|$ on the line is a century-old industry: the trivial *convexity* bound gives the exponent $\frac{1}{4}$, Weyl's *subconvexity* brought it to $\frac{1}{6}$, and successive efforts have pushed it down to about $\frac{13}{84}$ (Bourgain), creeping toward the Lindelöf value 0 without arriving. The same battle is fought over the moments $\int_0^T |\zeta(\frac{1}{2}+it)|^{2k}\,dt$, conjectured by Keating and Snaith—again from random matrices—to grow like $c_k\,T(\log T)^{k^2}$, proved only for $k \le 2$. A close cousin is the Dirichlet divisor problem: the error

$$\Delta(x) = \sum_{n \le x} d(n) - x\big(\log x + 2\gamma - 1\big) \tag{8.5}$$

is conjectured to be $O(x^{1/4+\varepsilon})$ and known only to about $x^{0.31}$, the very same kind of exponent that will not close. The verification computes $\Delta(x)$ to 10^6 and finds it comfortably $O(x^{0.4})$, its apparent exponent near 0.37. The lure is that subconvexity keeps improving and Lindelöf is the easier target; the turn is that the improvements slow to a crawl and never reach t^ε, and that even this lesser wall—RH's own shadow—has never been brought down. If the shadow resists, the substance is harder still.

8.4 The geometry wall: where the Hypothesis is a theorem

The last neighbour is the strangest, because there the Hypothesis is simply true. Replace the integers by the polynomials over a finite field $\mathbb{F}_q$ and the analogue of zeta is the zeta function of a curve C; its zeros are governed by the eigenvalues α of the Frobenius map, and the analogue of the Riemann Hypothesis is the statement

$$|\alpha| = q^{1/2}, \tag{8.6}$$

that the Frobenius eigenvalues lie on a single circle—the curve's critical line. This is the Weil conjecture for curves, proved by Weil in 1948 and in full generality by Deligne in 1974. For an elliptic curve it is exactly Hasse's bound $|a_p| \le 2\sqrt{p}$, with $a_p = p+1-\#E(\mathbb{F}_p)$, and the verification confirms it on nineteen curves; Figure 8.3 shows the Frobenius eigenvalues sitting precisely on the circle. The proof is a positivity—of the intersection form on the surface $C \times C$—and it is the same species of positivity the Weil criterion asks for over the integers.

For the elliptic curve that positivity is elementary, and worth seeing in full, because it is the whole climb in miniature. The Frobenius endomorphism $\phi\colon (x,y) \mapsto (x^p, y^p)$ sends E to itself and fixes exactly the points with coordinates in $\mathbb{F}_p$, so $\#E(\mathbb{F}_p) = \deg(1-\phi)$—which is what the count $a_p = p+1-\#E(\mathbb{F}_p)$ records. In the endomorphism ring ϕ obeys its characteristic relation

$$\phi^2 - a_p\,\phi + p = 0, \tag{8.7}$$

and the degree is a *positive-definite* quadratic form there—a nonzero endomorphism has positive degree—so for every pair of integers m, n,

$$\deg(m + n\phi) = m^2 + a_p\,mn + p\,n^2 \ \ge\ 0. \tag{8.8}$$

A binary form that never turns negative cannot have positive discriminant, and that discriminant is $a_p^2 - 4p$. Hence $a_p^2 \le 4p$—Hasse's bound $|a_p| \le 2\sqrt{p}$. The two roots

$\alpha, \beta = \frac{1}{2}(a_p \pm \sqrt{a_p^2 - 4p})$ are then a conjugate pair of absolute value $\sqrt{p}$: the Frobenius eigenvalues, on the circle that is the curve's critical line. The local Riemann Hypothesis, proved—by the elementary cousin of the intersection-form positivity that handles the general curve.

It is worth doing once by hand. Take $E\colon y^2 = x^3 + x + 1$ over $\mathbb{F}_7$. Running x through $0, \ldots, 6$ and keeping those for which $x^3 + x + 1$ is a square leaves just two, $x = 0$ and $x = 2$, each with two values of y; with the point at infinity that is five points in all, $\mathcal{O}$, $(0, \pm 1)$, $(2, \pm 2)$. So $\#E(\mathbb{F}_7) = 5$ and $a_7 = 7 + 1 - 5 = 3$, and the bound reads $3 \le 2\sqrt{7} \approx 5.29$, met with room. The characteristic relation $\phi^2 - 3\phi + 7 = 0$ has roots $\frac{1}{2}(3 \pm \sqrt{-19})$ of absolute value $\sqrt{(9+19)/4} = \sqrt{7}$; the Frobenius eigenvalues sit on the circle of radius $\sqrt{7}$ at angle $\theta_7 = \arccos(\frac{3}{2\sqrt{7}}) \approx 55.5°$. One curve by hand; the figure does nineteen at once.

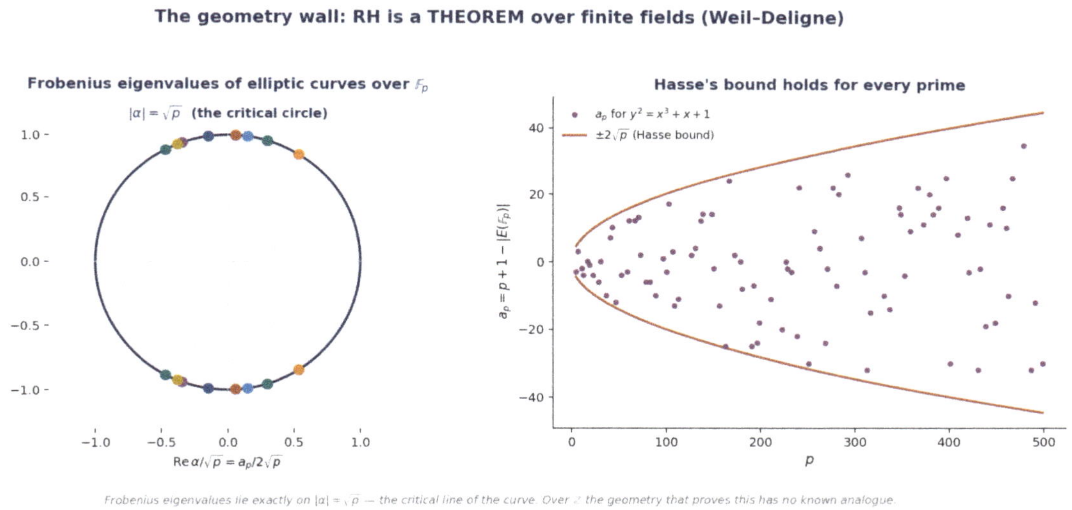

Figure 8.3: The geometry wall. *Left:* the (normalized) Frobenius eigenvalues of elliptic curves over $\mathbb{F}_p$ lie exactly on the unit circle $|\alpha| = \sqrt{p}$—the curve's critical line. *Right:* Hasse's bound $|a_p| \le 2\sqrt{p}$ holds for every prime. The function-field Riemann Hypothesis is a theorem; over $\mathbb{Z}$ the geometry that proves it has no known analogue.

The lure is overwhelming: the Hypothesis is proven next door, by a method that looks built to be carried home, and the positivity it uses is the one the integers need. The dream is to find a geometry for $\operatorname{Spec}\mathbb{Z}$—a "curve" over the elusive field with one element $\mathbb{F}_1$ (Manin, Deninger, Connes–Consani)—on which the same argument would run. The turn is that the geometry is not there. $\operatorname{Spec}\mathbb{Z}$ is not a curve over a field; there is no surface $\mathbb{Z} \times \mathbb{Z}$, no Frobenius, no $\mathbb{F}_1$ that has been made to carry the proof. The tools that win in the function-field world—the very tools that proved the local Hypothesis for Ramanujan's τ one chapter back—have no counterpart over the integers. The Hypothesis is a theorem in the next world and a wall in this one, for want of a geometry.

8.5 The ring

Stand back and the four neighbours close into a ring (Figure 8.4). The spacing statistics say an operator should exist; the operator, if it existed, would give the Hypothesis for free and would supply the Weil positivity of the positivity face; the function-field world possesses exactly the geometry that yields that positivity and proves the Hypothesis there; and yet the geometry, the operator, and therefore the positivity are all missing over the integers. Lindelöf, the weakest of the four, still stands unproved beside them. Each neighbour points inward at the wall and sideways at the others, and none reaches the centre.

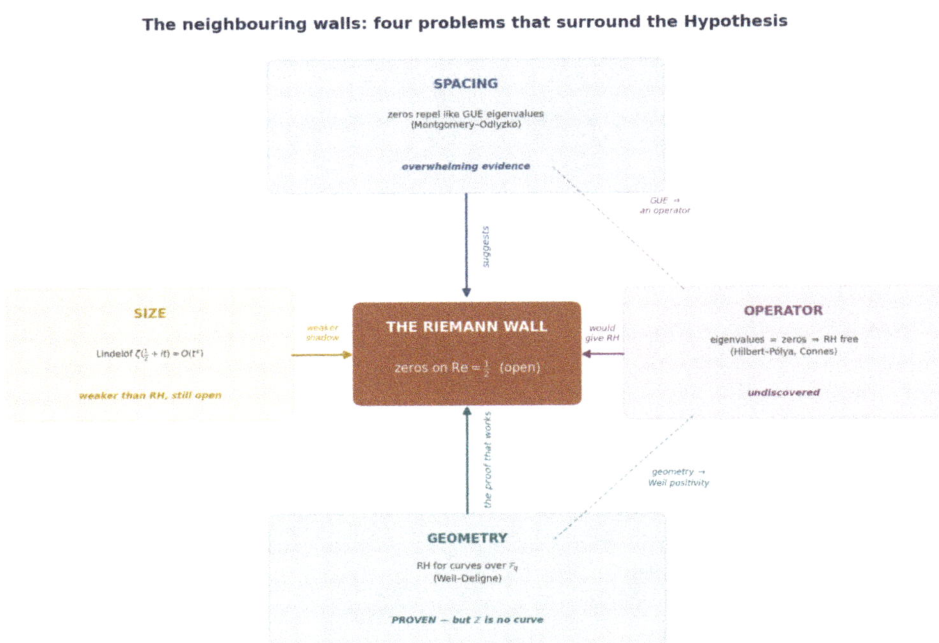

Figure 8.4: The neighbouring walls. Spacing (overwhelming evidence), the operator (undiscovered), size (weaker than RH, still open), and geometry (proven, but $\mathbb{Z}$ is no curve) surround the central wall. The arrows are real implications and analogies; the centre is what none of them reaches.

What this is and is not

What is genuine here is a great deal, more than on any single face: the Montgomery–Odlyzko statistics, exact and spectacularly confirmed; the subconvexity bounds and proven moments and the divisor estimates; the Hilbert–Pólya, Berry–Keating, and Connes programmes, each a real piece of mathematics; and above all the function-field Hypothesis, a full theorem of Weil and Deligne, verified here on actual curves. The verification exhibits the repulsion of the zeros, the spectral staircase $N(T)$, the small divisor error, and Hasse's bound holding exactly.

What it is not is a proof, on any of the four sides. The statistics are a resemblance, not a theorem about a single zero; the operator the statistics demand has never been constructed; the size bound that is weaker than the Hypothesis is itself open; and the geometry that proves the Hypothesis next door has no analogue over the integers. These are the most exciting things in the neighbourhood of the wall, and they are exactly the measure of its isolation: a wall ringed by proven cousins, an all-but-certain operator, and its own unproven shadow, with open ground between every one of them and the centre.

Verification

The script below, standard library only and deterministic, touches each neighbour: it unfolds the first sixty zeros and shows their GUE repulsion; checks the zero count against the Riemann–von Mangoldt staircase $N(T)$; computes the Dirichlet divisor error to 10^6 and finds it $O(x^{0.4})$; and confirms the function-field Hypothesis—Hasse's bound $|a_p| \leq 2\sqrt{p}$, Frobenius on $|\alpha| = \sqrt{p}$—on nineteen elliptic curves.

```
#!/usr/bin/env python3
"""Verification for 'The Neighbouring Walls'. Standard library only; deterministic."""
import math, time
t0=time.time()
PI=math.pi
checks=0
def ok(cond,msg):
    global checks
    assert cond,"FAIL: "+msg
    checks+=1; print("PASS:",msg)
GAMMA=[14.13472514,21.02203964,25.01085758,30.42487613,32.93506159,37.58617816,
 40.91871901,43.32707328,48.00515088,49.77383248,52.97032148,56.4462477,59.347044,
 60.83177852,65.11254405,67.07981053,69.54640171,72.06715767,75.7046907,77.14484007,
 79.33737502,82.91038085,84.73549298,87.42527461,88.80911121,92.49189927,94.65134404,
 95.87063423,98.83119422,101.31785101,103.72553804,105.44662305,107.16861118,
 111.02953554,111.87465918,114.32022092,116.22668032,118.79078287,121.370125,
 122.94682929,124.25681855,127.51668388,129.5787042,131.08768853,133.4977372,
 134.75650975,138.11604205,139.73620895,141.1237074,143.11184581,146.00098249,
 147.42276534,150.05352042,150.92525761,153.02469381,156.11290929,157.59759182,
 158.84998817,161.18896414,163.03070969]

# ===== SPACING WALL: the zeros repel like GUE eigenvalues =====
def Nsmooth(g): return (g/(2*PI))*math.log(g/(2*PI)) - g/(2*PI) + 7.0/8.0
u=[Nsmooth(g) for g in GAMMA]                # unfold by the mean density
sp=[u[i+1]-u[i] for i in range(len(u)-1)]
mean=sum(sp)/len(sp)
small=sum(1 for s in sp if s<0.5)/len(sp)      # fraction of tiny gaps
# GUE (Wigner) vs Poisson predictions for P(s<0.5):
poisson=1-math.exp(-0.5)                         # 0.393
print(f"  unfolded spacings: mean={mean:.3f}; frac(<0.5)={small:.3f} "
      f"(Poisson~{poisson:.3f}, GUE repels)")
ok(abs(mean-1.0)<0.05 and small<0.20,
   "zeros unfold to mean spacing 1 and repel (few small gaps): GUE, not Poisson")

# ===== OPERATOR WALL: the count is the Riemann-von Mangoldt staircase =====
T=GAMMA[-1]
predicted=Nsmooth(T)
print(f"  zeros up to T={T:.2f}: actual count=60, smooth N(T)={predicted:.3f}")
```

```
ok(abs(60-predicted)<1.5,
   "the zero count matches N(T)=(T/2pi)log(T/2pi)-T/2pi+7/8 (the spectral staircase)")

# ===== SIZE WALL: the Dirichlet divisor error stays small (Lindelof-type) =====
N=10**6
d=[0]*(N+1)
for i in range(1,N+1):
    for j in range(i,N+1,i): d[j]+=1
EG=0.5772156649015329
cum=0; maxabs=0.0; argmax=0
for x in range(1,N+1):
    cum+=d[x]
    Delta=cum-(x*(math.log(x)+2*EG-1))
    if abs(Delta)>maxabs: maxabs=abs(Delta); argmax=x
exp_emp=math.log(maxabs)/math.log(argmax)
print(f"  divisor error: max|Delta(x)|={maxabs:.1f} at x={argmax}; "
      f"apparent exponent={exp_emp:.3f}")
ok(maxabs < argmax**0.4,
   "Dirichlet divisor error Delta(x)=O(x^0.4) to 10^6 (conjectured x^(1/4+eps))")

# ===== GEOMETRY WALL: function-field RH is PROVEN (Hasse bound) =====
def legendre(a,p):
    a%=p
    if a==0: return 0
    return 1 if pow(a,(p-1)//2,p)==1 else -1
def ap(a,b,p):                                  # E: y^2 = x^3+ax+b over F_p
    return -sum(legendre((x*x*x+a*x+b)%p,p) for x in range(p))
worst=0.0; tested=0
for p in (11,13,101,103,1009):
    for (a,b) in ((1,1),(2,3),(0,7),(5,1)):
        if (4*a*a*a+27*b*b)%p==0: continue        # singular, skip
        A=ap(a,b,p);
        # Frobenius eigenvalues: roots of T^2 - A T + p; |alpha|=sqrt(p) iff A^2<=4p
        worst=max(worst, abs(A)/(2*math.sqrt(p)))
        tested+=1
print(f"  Hasse |a_p|<=2 sqrt(p): {tested} curves, worst ratio={worst:.4f}")
ok(worst<=1.0,
   "function-field RH is a THEOREM: |a_p|<=2 sqrt(p), Frobenius on |alpha|=sqrt(p)")

print(f"\nALL {checks} CHECKS PASSED   ({time.time()-t0:.1f}s)")
```

Chapter 9

The Irresistible Twin Corridor

No structure in the elementary theory pulls harder than the twin corridor. Fold a period of the wheel and the twins fall into eight clean lanes, each mirrored across the fold, each recurring as far as one cares to count; the empty stretches between them keep a fixed rhythm; and the first composite to break a corridor does so just where one might have guessed. The arrangement looks so plainly engineered that it seems it must be a key to the oldest question about it—whether the twins go on forever. This chapter follows that pull to its end. The corridors are real, the symmetry exact, the recurrence genuine—and every part of it is reflection and equidistribution against a single constant fixed once and for all. The lure is honest. It simply is not a lever, and the work of the chapter is to feel it completely and then say so.

Scope

Of all the channels that ride the wheel, the twin corridor is the one that most invites belief. Fold a single period of 210 at its midpoint and the twin primes fall into eight clean lanes, each mirrored across the fold, each recurring period after period as far as one cares to count; the empty stretches between the lanes keep a fixed rhythm; and the smallest composite that breaks a corridor does so at a place one could almost have predicted. The arrangement is so plainly ordered that it seems it must be a key to the oldest open question about it—whether the twins go on forever.

This chapter follows that pull deliberately, and to its end. The first half draws the corridor out in full and lets the order accumulate: the lanes, the reflection that folds fifteen corridors to eight, the fixed rhythm of the deserts, the recurrence across periods. The second half turns. The groupings the eye seizes on are reflection symmetry and equidistribution and nothing besides; the whole recurrence is one constant, the twin-prime constant, computed once and repeating; and the wall—the parity barrier—stands exactly where it has always stood. The aim is to feel the lure completely and then to name, precisely, why it is a lure and not a lever.

In the arc of this book this is the second of the three great walls, reached from the same base as the first. The wheel hands over the corridor and its density just as, in the opening chapter, it handed ζ over to the explicit formula—one foundation, climbed by a different route. And the wall the corridor reaches is not the twins' alone: the parity barrier that stops it is the same obstruction the Liouville sum meets on the oscillation face, the point past which a sieve can fix a pattern's density exactly and still never certify that the pattern recurs. The twin wall is climbed from the foundation of the opening chapter and turned back for a reason that rhymes, in the end, with the Riemann wall's own.

9.1 The corridors, drawn

Begin with the picture (Figure 9.1). Five periods of 210 stand side by side, and each is folded at its midpoint 105: the left column climbs $-1, 0, 1, \ldots$ up toward 105 while the right column descends $211, 210, 209, \ldots$ down to meet it, so that every row carries a residue r beside its reflection $210 - r$, with the dashed axis between the columns marking the fold. Twin primes are filled green; a prime with no twin is left plain; the composite spokes—the subprimes—are orange; the rim, the positions divisible by $2, 3, 5$ or 7, is faint.

The twins live only in the corridors. A residue r can hold a twin only if both r and $r + 2$ are coprime to 210, and there are exactly fifteen such residues—the fifteen twin corridors. No twin ever touches the rim, the positions divisible by $2, 3, 5$ or 7; the corridors are the only doors.

The fold halves the count. The reflection $r \mapsto 210 - r$ carries the corridor at r to the corridor at $208 - r$—the pair $(r, r + 2)$ reflects to $(208 - r, 210 - r)$—so the fifteen corridors pair off into seven mirrors, $11 \leftrightarrow 197$, $17 \leftrightarrow 191$, $29 \leftrightarrow 179$, $41 \leftrightarrow 167$, $59 \leftrightarrow 149$, $71 \leftrightarrow 137$, $101 \leftrightarrow 107$, together with one self-paired seam: the corridor $(209, 211)$ that straddles the boundary between periods, where $211 \equiv 1$ opens the next one, marked $209 \leftrightarrow 1$. Eight folded corridors in all, shaded and named down the left margin of the figure, and a twin in the left column sits at the very same height as a twin-admissible row in the right.

And the lanes recur. Read across the five periods and the same eight lanes light up again and again, the twins thinning as the numbers grow—roughly as $1/(\log x)^2$, the density falling but the lanes never moving. It is hard, looking at this, not to feel that the recurrence is doing something: that the lanes are rails the twins are obliged to run on, period after period, and that the obligation might somehow be made to yield the infinitude of twins. Hold that feeling. It is exactly the right feeling to have, and we will account for every part of it.

Figure 9.1: The twin corridors across five periods of 210. Each period is folded at 105: the left column runs $-1 \to 105$ and the right $211 \to 105$, so a residue r sits beside its reflection $210 - r$ across the dashed axis. Twin primes are green, lone primes plain, the subprimes (composite spokes) orange, the rim faint. The fifteen twin corridors fold to eight shaded, named lanes—seven mirror pairs $r \leftrightarrow 208 - r$ $(11 \leftrightarrow 197, \ldots, 101 \leftrightarrow 107)$ and the self-paired seam $(209, 211)$ shown as $209 \leftrightarrow 1$. In the $41 \leftrightarrow 167$ lane the genuine twin $(41, 43)$ stands on the left, while on the right the slot $(167, 169)$ carries the red twin-symmetry break $169 = 13^2$, so the twin the reflection would place there never forms. The lanes recur period to period and the twins thin; the alignments are reflection symmetry and equidistribution, not a correlation.

9.2 The rhythm of the deserts

Between the lanes lie the deserts—the runs of consecutive integers that hold no twin at all. In the lower half of the period they come in a fixed sequence of lengths: $3, 3, 9, 9, 15, 9, 27$. After the twin $(5, 7)$ three barren integers, then after $(11, 13)$ three more; then the deserts widen to nine, nine, fifteen, nine, and finally the great stretch of twenty-seven empty integers from 74 to 100 before the corridor at 101 opens again.

This rhythm is not a sample. It is fixed by where the admissible residues fall mod 210, and so it is identical in every period: the deserts of the thousandth period have the same lengths in the same order as the deserts of the first. What changes from period to period is only whether a given corridor, when it comes, holds two primes—the lanes and the deserts themselves are permanent furniture. The order is real, and it is the order of a fixed set of residues, drawn once.

The same corridors, read through the least prime factor, show the skeleton they ride on (Figure 9.2). Fold two periods and colour every cell by its smallest prime factor: the $2-3-2$ backbone of the rim, with $\div 5$ and $\div 7$ threaded through it, all of it reflection-symmetric across each fold because 210 is divisible by $2, 3, 5, 7$. The twins sit as green rungs on that backbone, the lone primes pale beside them, and the subprimes—least factor ≥ 11—stand out in purple, $169 = 13^2$ among them. The barren runs between the green rungs are the deserts, in their fixed lengths; and the last-digit columns flanking each period show the wheel's other mirror, $1 \leftrightarrow 9$ and $3 \leftrightarrow 7$, folding in exact step. Two reflections—the residue fold and the last-digit fold—run through one fixed skeleton.

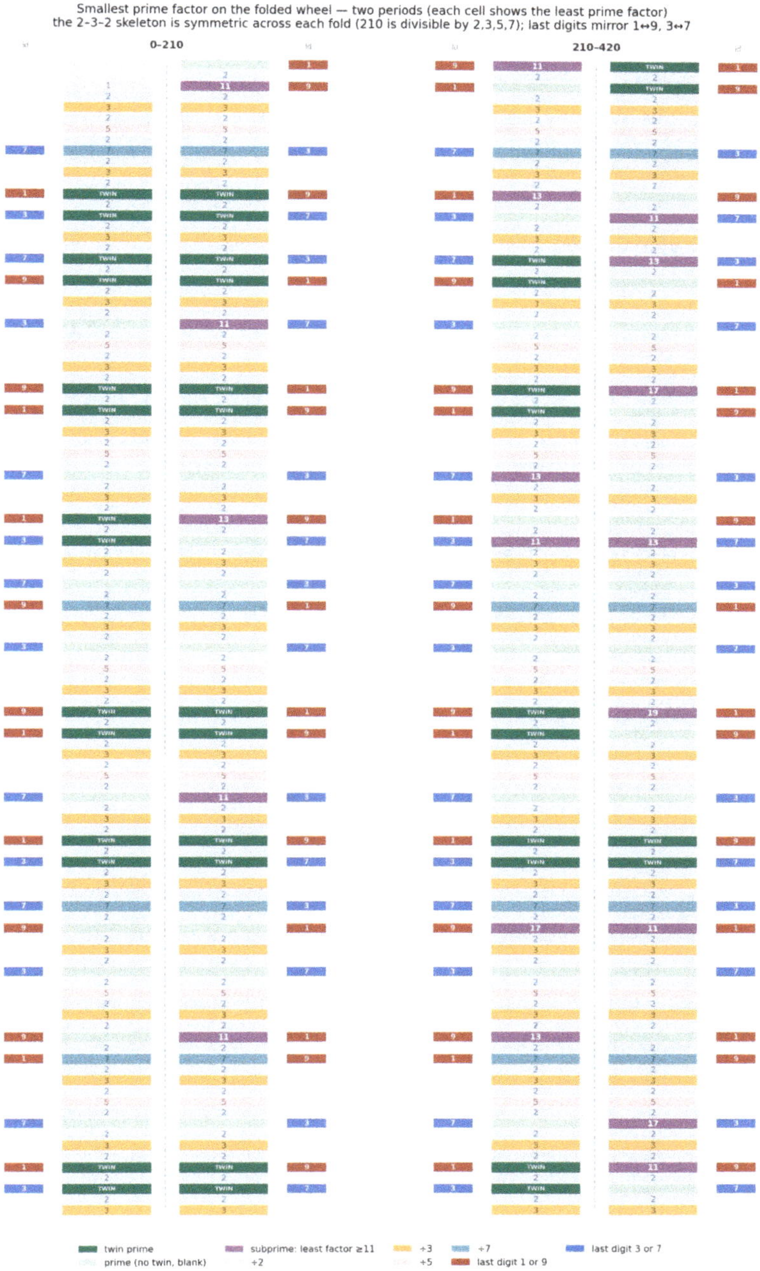

Figure 9.2: The same fold, read by least prime factor, over two periods. Each cell shows its smallest prime factor: ÷2 pale blue, ÷3 orange, ÷5 pink, ÷7 teal, and the subprimes (least factor ≥ 11) purple. The $2-3-2$ skeleton is reflection-symmetric across each fold and identical from one period to the next; twins are green, lone primes pale green, and the last-digit columns on either side mirror $1 \leftrightarrow 9$ and $3 \leftrightarrow 7$. The green twins ride the fixed backbone, the deserts are the barren runs between them, and the purple subprimes—$169 = 13^2$ the first to fall inside a corridor—multiply as the twins thin.

9.3 Where a corridor breaks from within

The reflection suggests a tidy expectation: if $(41, 43)$ is a genuine twin, its mirror corridor at 167 ought to be one too. It is not. The corridor is there—both 167 and 169 are coprime to 210, a legitimate twin-admissible slot—but $169 = 13^2$. The slot holds a prime and a square, not two primes. In the figure this is the lane where the left column is green twice over while the right carries a lone prime, 167, alongside the red break $169 = 13^2$.

This is the first place a composite breaks a twin corridor from within, and it happens exactly when the smallest prime outside the wheel, 13, first lands its square inside a corridor. It is the twin entry on the wheel's first-break ladder—11^2 for the bare primes, $11 \cdot 13$ for the cousins, 13^2 for the twins—and it is the first honest crack in the symmetry: the reflection organizes the corridors perfectly and says nothing whatever about whether a given corridor is filled by primes.

9.4 The groupings are equidistribution, not linkage

Now the turn. Spend long enough with the figure and the eye begins to find groupings: a corridor and its mirror both lit in one period, then both lit again in the next; clusters of green that seem to travel together. It is natural—irresistible, even—to read those alignments as evidence of a binding, some correlation by which corridors hold one another's twins in place.

They are not. Tested against the only honest baseline, a corridor and its mirror are co-realised—both holding a twin in the same period—no more often than any two corridors chosen at random. Across several thousand periods the mean correlation between mirror corridors and the mean correlation across all corridor pairs come out statistically the same, both indistinguishable from zero (the verification below reports $+0.002$ against -0.000). The alignments are the visible sum of two things already named and nothing else: the reflection symmetry, which is fixed, and the rough equidistribution of twins among the corridors, which is simply what equidistribution looks like. There is no third thing. The groupings carry no signal.

This is the hardest sentence in the chapter to believe while looking at the picture, which is exactly why it has to be said while looking at the picture.

9.5 What the corridor is made of

So what is the recurrence, if not a mechanism? It is a constant. The Hardy–Littlewood conjecture gives the expected number of twins up to x as

$$\pi_2(x) \sim 2C_2 \int_2^x \frac{dt}{(\log t)^2}, \qquad C_2 = \prod_{p \geq 3} \left(1 - \frac{1}{(p-1)^2}\right) = 0.6601618\ldots,$$

the twin-prime constant. The density of twins in the corridors is this single number, C_2, times the same Mertens thinning the primes already feel. The lanes that recur period after period, the rhythm of the deserts, the whole sense of an engineered structure—all of it is the singular series C_2, an Euler product over the primes computed once and then repeated.

That is the deflation in one line. The recurrence is not structure held in reserve to be spent on the twin question; it *is* the answer to a question already answered—how dense would twins be, if they are infinite—wearing the look of a lever on the question still open, whether they are. The Euler product is cashed. There is nothing behind the lanes but the constant that defines them.

9.6 The wall

The question the corridor cannot answer is the oldest one: are there infinitely many twins? The corridor says where they must live and how thickly they would lie, and stops. Brun proved that the sum of the reciprocals of the twins converges, so they are at most sparse; Hardy and Littlewood conjectured the exact density above; and there it has stood.

The genuine progress of the last decade did not come from the corridor. Zhang, and then Maynard and Tao, proved that infinitely many pairs of primes sit within a bounded gap—below seventy million at first (Zhang, 2013), reduced to 600 by Maynard and Tao and to 246 by the Polymath project—which is real passage into the dimension-two channel. But it came from sieve theory clever enough to bend the parity barrier, not from the wheel's geometry; the corridors neither suggested it nor are touched by it. Between *bounded gap* and *gap two* stands the parity barrier of Selberg, the same one unit of sieve dimension that separates the solved prime channel from the open twin channel—and the same barrier that, on the oscillation face, holds the Liouville sum just beyond a sieve's reach.

The corridor, then, is the wall's own description. It is an exact, complete, and beautiful account of where twins can be and how dense they are—which is to say, it is the singular series and the parity barrier, drawn as a picture. It is not, and never was, a way across.

What this is and is not

What is genuine here is all of it. The fifteen corridors, the eight folded lanes, the seven mirror pairs and the seam, the desert rhythm $3, 3, 9, 9, 15, 9, 27$, the break at 13^2, the constant $C_2 = 0.6601618\ldots$—every one is real, exact, and elementary, and most of it is older than a century. None of it is a trick.

What it is not is a key. The twin corridor is the most irresistible object in the elementary theory precisely because every part of its order is real: the symmetry does not lie, the recurrence does not lie, the deserts keep their rhythm honestly. The lure is not deception. It is the genuine pull of genuine structure, and the only error on offer is the last one—to mistake the structure for a mechanism, the singular series for a lever, the picture of the wall for a door in it. We have let the corridor pull as hard as it pulls and named, at each step, exactly what the pull is made of. That naming is the whole of the defence, and it is the only one there is.

Verification

Every count and constant in this chapter is closed-form or a direct, seed-free computation. The script below, using only the Python standard library, recomputes the fifteen corridors, the eight folded lanes, the desert rhythm, the 13^2 break, the twin-prime constant, and the absence of any mirror-corridor linkage, by assertion.

```
#!/usr/bin/env python3
"""Verification for 'The Irresistible Twin Corridor'.
Standard library only; deterministic (no random seeds)."""
from math import gcd

def sieve(n):
    s=bytearray([1])*(n+1); s[0]=s[1]=0
    i=2
    while i*i<=n:
        if s[i]:
            s[i*i::i]=bytearray(len(s[i*i::i]))
        i+=1
    return s

M=210
checks=0
def ok(cond,msg):
    global checks
    assert cond, "FAIL: "+msg
    checks+=1
    print("PASS:",msg)

# --- the fifteen twin corridors of the wheel ---
corridors=[r for r in range(M) if gcd(r,M)==1 and gcd((r+2)%M,M)==1]
ok(len(corridors)==15, f"twin corridors mod 210 number 15 (got {len(corridors)})")

# --- the fold r <-> 208-r : seven mirror pairs + one self-paired seam = eight lanes ---
S=set(corridors); lanes=set(); selfpaired=[]
for r in corridors:
    m=(208-r)%M
    if m==r: selfpaired.append(r)
    lanes.add(frozenset({r,m}))
ok(len(lanes)==8, f"the fifteen fold to eight lanes (got {len(lanes)})")
pairs=[l for l in lanes if len(l)==2]
ok(len(pairs)==7 and len(selfpaired)==1,
   f"seven mirror pairs + one seam (got {len(pairs)} pairs, seam {selfpaired})")
ok(selfpaired==[209], f"the seam is the corridor 209 (got {selfpaired})")

# --- the rhythm of the deserts in the lower half ---
sm=sieve(220)
tp=[p for p in range(3,104) if sm[p] and sm[p+2]]      # twin lower-members in [1,105]
gaps=[]
for (a,b) in zip(tp,tp[1:]):
    g=b-a-3             # integers strictly between a+2 and b
    if g>0: gaps.append(g)
ok(gaps==[3,3,9,9,15,9,27], f"twin-desert rhythm 3,3,9,9,15,9,27 (got {gaps})")

# --- the first break: 169 = 13^2 breaks the mirror of the twin (41,43) ---
ok(sm[167] and not sm[169] and 169==13*13, "167 prime, 169 = 13^2 composite")
```

```
ok(gcd(167,M)==1 and gcd(169,M)==1, "(167,169) is a genuine corridor (both coprime to 210)")
ok((208-41)%M==167, "corridor 167 is the mirror of corridor 41")
ok(sm[41] and sm[43], "(41,43) is a true twin; its mirror is broken from within")

# --- the constant the corridor is made of: the twin-prime constant C2 ---
sp=sieve(200000); primes=[p for p in range(3,200001) if sp[p]]
C2=1.0
for p in primes: C2*=1-1.0/((p-1)**2)
ok(abs(C2-0.6601618)<1e-4, f"twin-prime constant C2 = {C2:.7f} (~0.6601618)")

# --- the groupings are equidistribution, not linkage ---
# co-realisation of mirror corridor pairs vs. arbitrary corridor pairs, over many periods
N=4000; big=sieve(M*N+10)
ind={c:[1 if (big[M*k+c] and big[M*k+c+2]) else 0 for k in range(1,N)] for c in corridors}
def corr(a,b):
    n=len(a); ma=sum(a)/n; mb=sum(b)/n
    cov=sum((x-ma)*(y-mb) for x,y in zip(a,b))/n
    va=sum((x-ma)**2 for x in a)/n; vb=sum((y-mb)**2 for y in b)/n
    return cov/((va*vb)**0.5) if va>0 and vb>0 else 0.0
mirror=[(r,(208-r)%M) for r in corridors if r<(208-r)%M]
mc=sum(corr(ind[a],ind[b]) for a,b in mirror)/len(mirror)
allp=[(corridors[i],corridors[j]) for i in range(15) for j in range(i+1,15)]
ac=sum(corr(ind[a],ind[b]) for a,b in allp)/len(allp)
print(f"  mean correlation, mirror pairs : {mc:+.4f}")
print(f"  mean correlation, all pairs    : {ac:+.4f}")
ok(abs(mc-ac)<0.03 and abs(mc)<0.05,
   "mirror corridors co-realised no more than arbitrary pairs (no linkage)")

print(f"\nALL {checks} CHECKS PASSED")
```

Chapter 10

The Walls

This book has climbed a dozen methods, and met, each time, the same kind of wall: the place where the wheel's local law stops deciding the global fact. This closing chapter sets the three great walls side by side—the Riemann Hypothesis, the twin primes, and factoring—and says, for each, exactly what is known and exactly where the barrier stands. They are three faces of one situation. The primes give up their local structure freely—their residues, their densities, the singular series, the characters—and their global structure not at all: where the zeros lie, whether two primes recur forever, how to undo a product. What is known is genuine and often deep, and it was won by the very methods that then disclosed their own limits. What is not known is not for want of trying. At each wall the lever that would cross it turns out to be the thing it was trying to find.

Scope

This is a survey, not a new climb. The chapters before it each took one method from the wheel to a wall; here we stand back and look at the walls themselves. Three of them carry the weight of the subject. The Riemann Hypothesis is the wall of *location*: where, exactly, the zeros sit, and so how regular the primes are. The twin primes are the wall of *recurrence*: whether a pattern whose density we know exactly occurs forever. Factoring is the wall of *decomposition*: whether the primes, so easy to multiply together, can be pulled back apart. The three are distinct theorems with distinct barriers, and yet one situation runs through all of them. We will name, for each, what is proven, and then the one thing that is not, and why the not is so hard.

10.1 What the wheel gave, and where it stops

Look back along the climb and a single shape repeats. Each chapter took the wheel's local data and cashed it for a genuine, finished fact. The opening chapter cashed the Euler product and the family of L-functions $L(s,\chi)$—with the non-vanishing $L(1,\chi) \neq 0$ that carries Dirichlet's theorem—in exact closed form. The Ramanujan chapter cashed the Sato–Tate law and Deligne's bound, the proven local Hypothesis one floor up. The twin corridor cashed the singular series, the twin constant $2C_2 = 1.3203\ldots$, the local product that predicts the twin density to within a percent. And the five faces each cashed a local equivalent of the Hypothesis itself—an error term, a positivity, a completeness, an inequality, a heat-flow constant—while the neighbouring walls cashed the random-matrix statistics of the zeros and the proven Hypothesis over finite fields. Every one of these is local in the deepest sense: a statement about residues modulo the primes, assembled into a product, a count, or a transform, and then it is done.

And every one of them stopped at the same kind of edge. The local law gives the main term and falls silent on the fluctuation; gives the density and falls silent on the occurrence; gives the sieve and falls silent on the factorization. The wheel is the primes' local law, and the three walls are the three places that law runs out. Figure 10.1 sets them side by side with what is known on each.

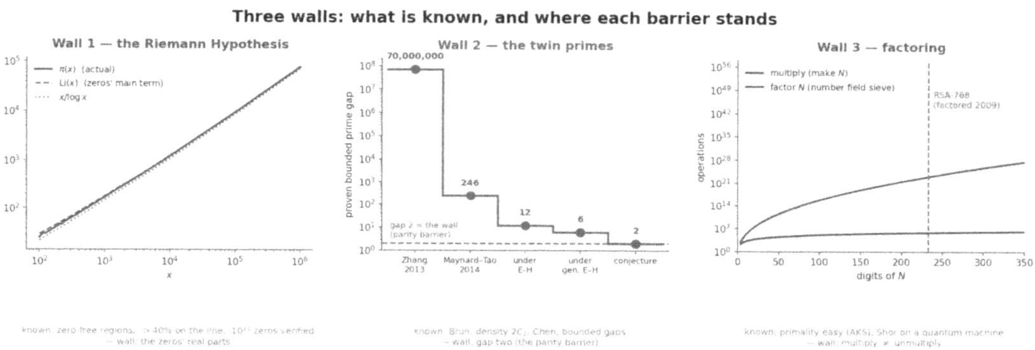

Figure 10.1: The three walls, with what is known on each. *Left:* for the Riemann Hypothesis, $\mathrm{Li}(x)$ tracks $\pi(x)$ far more closely than $x/\log x$—the better main term the zeros supply—and the residual error is what the Hypothesis is about. *Centre:* for the twins, the proven bound on prime gaps has fallen from Zhang's 70,000,000 to 246, and to 6 under a stronger hypothesis, but stops one step above gap two. *Right:* for factoring, making a number by multiplication is cheap while unmaking it by the number field sieve is not, the gap widening without bound in the number of digits.

10.2 The first wall: the Riemann Hypothesis

The Hypothesis is the wall of location: that every nontrivial zero has real part exactly $\frac{1}{2}$. By the explicit formula, this is the same as saying the primes are as regular as arithmetic permits—that the error in the prime number theorem is as small as it can possibly be. A great deal is known. Riemann's functional equation pins the zeros into a symmetric strip; de la Vallée Poussin's zero-free region, widened by Vinogradov and Korobov, is exactly what proves the prime number theorem with its error term; Selberg, then Levinson, then Conrey showed that a positive proportion—more than forty percent—of the zeros lie on the line; and the first many trillions of them have been checked, every one on the line, by direct computation. There are even fully elementary equivalents: Lagarias showed that the single inequality $\sigma(n) \le H_n + e^{H_n} \log H_n$, for every n, is equivalent to the Hypothesis—and the verification confirms it holds throughout its range. The faces of the earlier chapters are all this one wall seen from different sides—the oscillation of the error term, the positivity of a functional, the completeness of a span, an extremal inequality, the constant of a heat flow—each provably equivalent to the Hypothesis and none of them a way past it; and the neighbouring walls ring it with a proven analogue over finite fields, an all-but-certain operator, and its own weaker shadow.

What is missing is any control of the real parts themselves. The zero-free regions creep toward the line and never reach it; the proportion on the line climbs and never reaches one; the computations verify and never conclude. The reason is the one this book has met at every turn: the dependence runs from the zeros to the primes, never back. The explicit formula expresses the primes through the zeros, so everything we know about the primes is downstream of the zeros, and nothing we know about the primes can reach up to constrain where the zeros are. The Hypothesis is the cleanest statement of the wall: we can describe the consequences of the zeros' location to any precision, and we cannot touch the location.

10.3 The second wall: the twins

The twin primes are the wall of recurrence, and here the book's own constant stands at the foot. We know the twins are sparse—Brun proved that the sum of their reciprocals converges—and we know exactly how sparse they would be: the density is the singular series $2C_2$, drawn from the wheel, matching the counts to a percent. We know more than density. Chen proved that infinitely often p and $p+2$ are a prime and a number with at most two prime factors—a near-twin, missing by the thinnest possible margin. And in the last decade the gap itself was bounded: Goldston, Pintz and Yıldırım opened the door, Zhang walked through it with a bound of seventy million, and Maynard and Tao, with the Polymath effort, brought it down to 246, to 12 under the Elliott–Halberstam conjecture, and to 6 under its generalized form. Infinitely many primes within a bounded gap is a theorem.

But the last step does not come. Between *bounded gap* and *gap exactly two* stands the parity barrier of Selberg—the same barrier the twin corridor met at its summit. A sieve cannot tell an even number of prime factors from an odd one; it cannot separate "two primes" from "a prime and a product of two primes," which is exactly Chen's almost-twin. The verification shows the blindness directly: the signed parity sum, the Liouville function over the first million integers, is tiny relative to its range, a quantity the sieve cannot see and cannot use. Chen's theorem is the sieve pressed to its very edge. Gap two is one parity beyond the edge.

10.4 The third wall: factoring

Factoring is the wall of decomposition, and it is a different kind of wall—not a barrier to proof but a barrier to computation, the one the modern world is built on. Its shape is an asymmetry. To *make* a composite is trivial: multiply two primes, an instant's work.

To recognise a prime is also easy now—the algorithm of Agrawal, Kayal and Saxena tests primality in polynomial time, and fast probabilistic tests predate it by decades. But to *unmake* a composite, to recover the primes from their product, the best classical method known, the number field sieve, takes time that grows sub-exponentially in the digits, fast enough that a number of a few hundred digits is beyond reach. Shor showed that a quantum computer would factor in polynomial time, so the wall is not absolute—but classically it stands, and public-key cryptography rests on it.

The wheel sees the asymmetry and cannot cross it. Its sieve strikes out every multiple of $2, 3, 5, 7$ at a glance, and identifies the subprimes—the composites coprime to 210, whose least factor is at least eleven—as the survivors. But it cannot factor a subprime; it knows only that the small primes are absent. The verification makes the point concrete: take two primes near ten million, multiply them, and the product has no factor below a hundred thousand, so the wheel's sieve learns nothing, while the product itself was formed in a single multiplication. Easy to make, hard to unmake. The primes are the multiplicative atoms, and the wall is that atoms combine freely and split only under great force.

10.5 One wall, three faces

Set the three together and the same situation shows three times (Figure 10.2). In each, the wheel— the primes' local law—is known and sits at the foot; in each, a global fact stands above it, walled; and in each, the barrier is that the lever which would cross is the very thing being sought. To cross the Riemann wall one would need to control the zeros' real parts, but the zeros are what govern the primes one is reasoning from. To cross the twin wall one would need to break parity, but breaking parity is precisely what isolating two primes means. To cross the factoring wall cheaply one would need to undo multiplication, but undoing multiplication is precisely factoring. The destination is the tool. That circularity is not a failure of ingenuity; it is the structure of the subject.

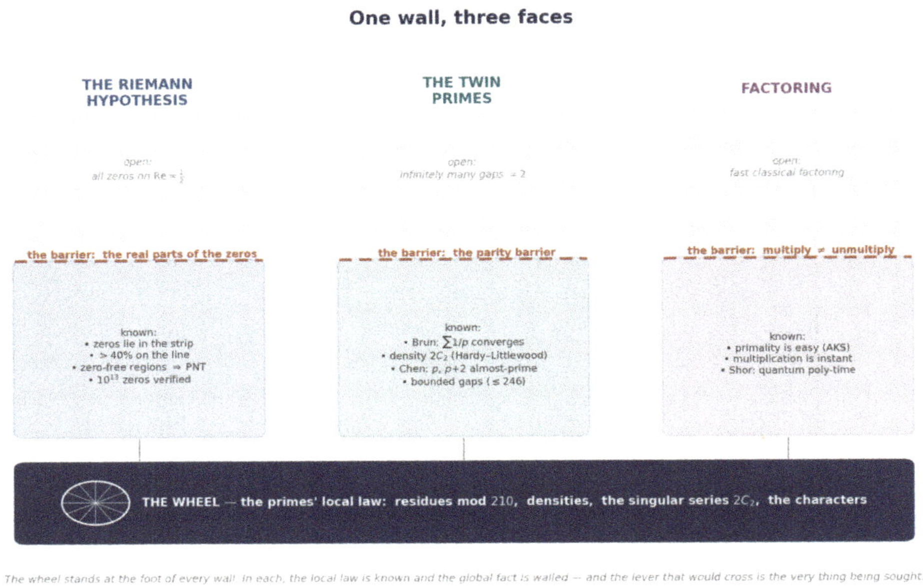

Figure 10.2: One situation, three faces. The wheel—residues, densities, the singular series, the characters—is the known foundation under every wall. On each, the local law is settled (below the barrier) and the global fact is open (above it): the real parts of the zeros, the parity barrier, the multiply–unmultiply asymmetry. In each the lever that would cross is the thing being sought.

10.6 What scaling the wall has meant

We have not crossed. That was never the promise. The promise was to climb each wall honestly, with the wheel at its foot, feeling the full pull of each method that seemed for a moment to lead over—and to name, at the top, exactly what the pull was made of. The gains along the way are real and were not small: the singular series and the corridor it draws; the non-vanishing of the L-functions in closed form; a positive proportion of the zeros on the line; primes recurring within a bounded gap; primality made easy; the factoring asymmetry on which a global infrastructure safely rests. Every one of these was won by a method that, pressed further, showed its own limit—the lure and the wall in the same hand.

That is the whole of the book's argument, and the whole of its honesty. The structures are genuine: the symmetry does not lie, the recurrence does not lie, the statistics of the zeros are real to every checked decimal. The only error ever on offer was the last one—to mistake a description of the wall for a door in it, the singular series for a lever, the spectral fingerprint for a hand, the local law for the global fact. We have let each wall pull as hard as it pulls, granted it every genuine thing it had to give, and located, each time, the one thing it lacked. The wheel turns at the foot of the wall. The wall still stands. And to know precisely where it stands, and precisely why, is what a century and more of the best mathematics has bought—which is not a crossing, and is not nothing.

What this is and is not

What is genuine here is the entire ledger of the known. The zero-free regions and the prime number theorem; the proportion of zeros on the line; Lagarias's elementary equivalent; Brun's convergence, the twin constant $2C_2$, Chen's near-twins, the bounded gaps of Zhang and of Maynard and Tao; the easy primality of Agrawal, Kayal and Saxena, and Shor's quantum factoring. Every item is proven, named, and real. None of it is a trick.

What it is not is a way over any of the three. The Riemann Hypothesis, the infinitude of twins, and fast classical factoring are open, and the verification marks the line precisely: an RH-equivalent inequality that holds as far as we look and is unproven beyond; a parity the sieve cannot see; a product cheap to form and costly to break. The three walls are three faces of the primes' refusal to hand over globally what they hand over locally for free. We have scaled each, with the wheel at its foot, and reported the view from the top without embellishment. That report is the book.

Verification

The known facts and the exact barriers are checked, where they can be made elementary, by the script below— standard library only, deterministic. It confirms an RH-equivalent inequality throughout its range, the better main term the zeros supply, the parity a sieve is blind to, and the make/unmake asymmetry of factoring.

```
#!/usr/bin/env python3
"""Verification for 'The Walls'. Standard library only; deterministic (no seeds)."""
from math import log, exp, isqrt

checks=0
def ok(cond,msg):
    global checks
    assert cond,"FAIL: "+msg
    checks+=1; print("PASS:",msg)

def sieve(n):
    s=bytearray([1])*(n+1); s[0]=s[1]=0
    i=2
    while i*i<=n:
        if s[i]: s[i*i::i]=bytearray(len(s[i*i::i]))
        i+=1
    return s

# ================= WALL 1: the Riemann Hypothesis =================
# (a) Lagarias's elementary inequality  sigma(n) <= H_n + exp(H_n) ln(H_n)
#     is EQUIVALENT to RH (Lagarias 2002). It holds throughout this range.
NL=100000
sigma=[0]*(NL+1)
for d in range(1,NL+1):
    for m in range(d,NL+1,d):
        sigma[m]+=d
H=0.0; lag=True; worst=0
```

```
for n in range(1,NL+1):
    H+=1.0/n
    rhs = 1.0 if n==1 else H+exp(H)*log(H)
    if sigma[n] > rhs+1e-9: lag=False; worst=n; break
ok(lag, f"Lagarias inequality holds for 1<=n<={NL} (RH-equivalent, unproven)")

# (b) the zeros' contribution: Li(x) tracks pi(x) far better than x/log x
X=1000000
pi=sum(sieve(X))
def Li(x):
    steps=200000; a=2.0; b=float(x); h=(b-a)/steps; tot=0.0
    for i in range(steps+1):
        w=0.5 if (i==0 or i==steps) else 1.0
        tot+=w/log(a+i*h)
    return tot*h
li=Li(X); rough=X/log(X)
print(f"  pi(10^6)={pi}; Li={li:.0f} (err {li-pi:+.0f})")
print(f"  x/logx={rough:.0f} (err {rough-pi:+.0f})")
ok(abs(li-pi) < abs(rough-pi)/10, "Li(x) beats x/log x (the zeros' main term)")

# ================= WALL 2: the twins (the parity barrier) =================
# A sieve cannot see the sign of Liouville lambda(n)=(-1)^Omega(n): cannot tell
# even from an odd number of prime factors -- exactly what separates a prime from a
# product of two primes. That blindness is the parity wall; the signed sum is tiny.
M=1000000
spf=list(range(M+1))
for i in range(2,isqrt(M)+1):
    if spf[i]==i:
        for j in range(i*i,M+1,i):
            if spf[j]==j: spf[j]=i
Om=[0]*(M+1)
for n in range(2,M+1):
    Om[n]=Om[n//spf[n]]+1
L=sum(-1 if Om[n]&1 else 1 for n in range(1,M+1))
print(f"  Liouville sum L(10^6) = {L}")
ok(abs(L) < 0.005*M, f"|L(10^6)|={abs(L)} small: the parity a sieve cannot see")

# ================= WALL 3: factoring (easy to make, hard to unmake) =================
def mr(n):
    if n<2: return False
    small=(2,3,5,7,11,13,17,19,23,29,31,37)
    for p in small:
        if n%p==0: return n==p
    d=n-1; s=0
    while d%2==0: d//=2; s+=1
    for a in small:
        x=pow(a,d,n)
        if x==1 or x==n-1: continue
        for _ in range(s-1):
            x=x*x%n
            if x==n-1: break
        else: return False
    return True
def nextprime(n):
    n+=1
    while not mr(n): n+=1
    return n
p=nextprime(10**7); q=nextprime(2*10**7)
N=p*q
ok(mr(p) and mr(q), f"primality is easy: {p} and {q} certified prime at once")
B=100000
ok(all(N%d!=0 for d in range(2,B+1)), f"sieve to {B} finds no factor of N (blind)")
ok(p*q==N and N>10**14, "N=pq is one multiply: easy to make, hard to unmake")
```

```
print(f"\nALL {checks} CHECKS PASSED")
```

Appendix A

Reformulations and Equivalences of the Riemann Hypothesis

The Riemann Hypothesis has been restated more ways than almost any open problem in mathematics, and the standing temptation is to file every restatement under one heading: equivalence. *They are not all equivalences. A clean catalogue has to keep three things apart—statements proven equivalent to the Hypothesis, approaches that would imply it or already assume it, and analogies that resemble it in another world—because conflating them is exactly the lever error the rest of this book has worked to name. A description is not a door; a hope is not a theorem; an analogy is not an identity. This appendix sorts the known reformulations by status, gives the sharp form of each genuine equivalence, flags two celebrated sharpenings that are simply false, and corrects two items that are not equivalences at all.*

Scope

This is a reference, sorted by what is actually known. The first part lists the genuine equivalences—the statements proven to hold if and only if every nontrivial zero has real part $\frac{1}{2}$—grouped by the kind of mathematics they come from. The second part records two famous strengthenings, of Mertens and of Pólya, that hold as far as anyone has computed and are nonetheless false: the sharpest reminders the subject offers that agreement in range is not truth. The third part places the spectral programme—Hilbert–Pólya, the random-matrix statistics, Berry–Keating, Connes—where it belongs, as approaches rather than equivalences; this is the spine of the companion volume, and is listed here only to be correctly labelled. The last part corrects two entries often miscatalogued: the function-field analogy, which is a different theorem in a different world, and the divisor bound, which is weaker than what is known unconditionally and so equivalent to nothing.

A.1 A map of the usual list

The fifteen statements most often gathered under "equivalences of RH" fall into the categories below.

#	Statement, as usually listed	Status
1	$\pi(x) = \mathrm{Li}(x) + O(x^{1/2+\varepsilon})$	equivalence
2	$\psi(x) = x + O(x^{1/2+\varepsilon})$	equivalence
3	Mertens $M(x) = O(x^{1/2+\varepsilon})$	equivalence
4	Liouville $L(x) = O(x^{1/2+\varepsilon})$	equivalence
5	Li's criterion, $\lambda_n \geq 0$ for all n	equivalence
6	Nyman–Beurling, 1 in an L^2 closure	equivalence
7	Hilbert–Pólya self-adjoint operator	would imply RH
8	Riemann ξ in the Laguerre–Pólya class	equivalence
9	Random-matrix (GUE) pair correlation	assumes RH; evidential
10	Berry–Keating $H = xp$	conjectural model
11	Divisor bound $O(x^{1/2+\varepsilon})$	*not* equivalent (trivial)
12	Möbius exponential sum	nuanced (see text)
13	Connes' noncommutative geometry	programme
14	Weil conjectures over finite fields	analogy
15	Quantum-chaos correspondence	analogy

A.2 The genuine equivalences

The error-term family (1–4). These four are one phenomenon—square-root cancellation—in four costumes. The prime counting function against the logarithmic integral, Chebyshev's ψ against x, the Mertens sum $M(x) = \sum_{n\leq x} \mu(n)$, and the Liouville sum $L(x) = \sum_{n\leq x} \lambda(n)$: each being $O(x^{1/2+\varepsilon})$ is equivalent to the Hypothesis. The sharp forms are due to Schoenfeld: on RH, $|\pi(x) - \mathrm{Li}(x)| < \frac{1}{8\pi}\sqrt{x}\log x$ for $x \geq 2657$, and $|\psi(x) - x| < \frac{1}{8\pi}\sqrt{x}(\log x)^2$. The verification confirms that $M(x)$ and $L(x)$ stay well inside an $x^{0.6}$ envelope to 10^6.

The function-theoretic family (5, 8). Li's criterion makes RH a positivity: the numbers $\lambda_n = \sum_\rho \bigl(1 - (1 - 1/\rho)^n\bigr)$ are all ≥ 0 exactly when the Hypothesis holds. Membership of the Riemann ξ in the Laguerre–Pólya class is RH almost by definition—that class is the closure of real polynomials with only real zeros, so ξ lies in it precisely when its zeros are all real. The modern sharpening is the de Bruijn–Newman constant Λ: RH is equivalent to $\Lambda \leq 0$, and the inequality $\Lambda \geq 0$ was proven by Rodgers and Tao in 2020, so RH is now equivalent to $\Lambda = 0$—the Hypothesis, if true, is true by the thinnest possible margin.

The Hilbert-space family (6). The Nyman–Beurling criterion makes RH an approximation problem: the constant function 1 lies in the $L^2(0,1)$-closure of the span of the dilated fractional parts $\{x/n\}$ exactly when the Hypothesis holds. Báez-Duarte's refinement (2003) sharpened this to a single, cleaner approximating sequence and is the form most used today.

The arithmetic family (additions). Two elementary inequalities deserve a place the usual list omits. Robin's inequality (1984): for every $n > 5040$, $\sigma(n) < e^{\gamma} n \log\log n$, and this holds for all such n if and only if RH holds; the verification checks it past its last exception, throughout $5040 < n \leq 10^6$. And Lagarias's inequality (2002), $\sigma(n) \leq H_n + e^{H_n} \log H_n$ for every n, with equality only at $n = 1$, likewise equivalent to RH—this is the one cashed in the Walls chapter. To these one may add the Riesz criterion, the Franel–Landau equidistribution of Farey fractions, and the Redheffer matrix, whose determinant equals $M(n)$.

Positivity (the genuine core of the spectral programme). Weil's explicit-formula criterion: RH holds if and only if a certain quadratic functional, built from the explicit formula, is positive. This is a true equivalence, and it is the part of the Connes programme that is settled mathematics rather than aspiration.

The fractal-geometric family (an addition the usual list omits). A genuine equivalence reaches the catalogue from a corner the standard fifteen rarely touch: the geometry of fractal strings, due to Lapidus and his collaborators. A *fractal string* is nothing more than a bounded open subset of the line, recorded by the decreasing sequence of its interval lengths $\ell_1 \geq \ell_2 \geq \cdots$; its *geometric zeta function* is $\zeta_{\mathcal{L}}(s) = \sum_j \ell_j^s$, and the poles of that function are its *complex dimensions*. The Cantor string—the gaps of the middle-thirds set, one length 1/3, two of 1/9, four of 1/27—has $\zeta_{\mathcal{L}}(s) = 1/(3^s - 2)$, whose poles sit at $s = \frac{\log 2}{\log 3} + i\,n\,\frac{2\pi}{\log 3}$: a vertical lattice whose real part is the set's fractal dimension and whose imaginary parts encode the log-periodic oscillations of its geometry (Figure A.1). The complex dimensions are to a fractal's geometry what the zeros are to the primes—the frequencies of an explicit ("tube") formula for its volume. The Riemann zeta enters because a string can be *heard*: the frequencies of its intervals are k/ℓ_j, and their spectral zeta function factors as $\zeta(s)\,\zeta_{\mathcal{L}}(s)$, making ζ the exact bridge from geometry to spectrum. On that bridge Lapidus and Maier proved the equivalence: the inverse spectral problem—whether a string's spectrum fixes its Minkowski geometry—has an affirmative answer for every dimension $D \in (0, 1)$ *except* $D = \frac{1}{2}$ if and only if the Riemann Hypothesis holds. The exceptional dimension is $\frac{1}{2}$ for the obvious reason, and RH becomes the statement that $\frac{1}{2}$ is the single dimension at which one cannot hear a fractal string's shape. Like every entry in this section it relocates the wall—here into fractal geometry—without lowering it.

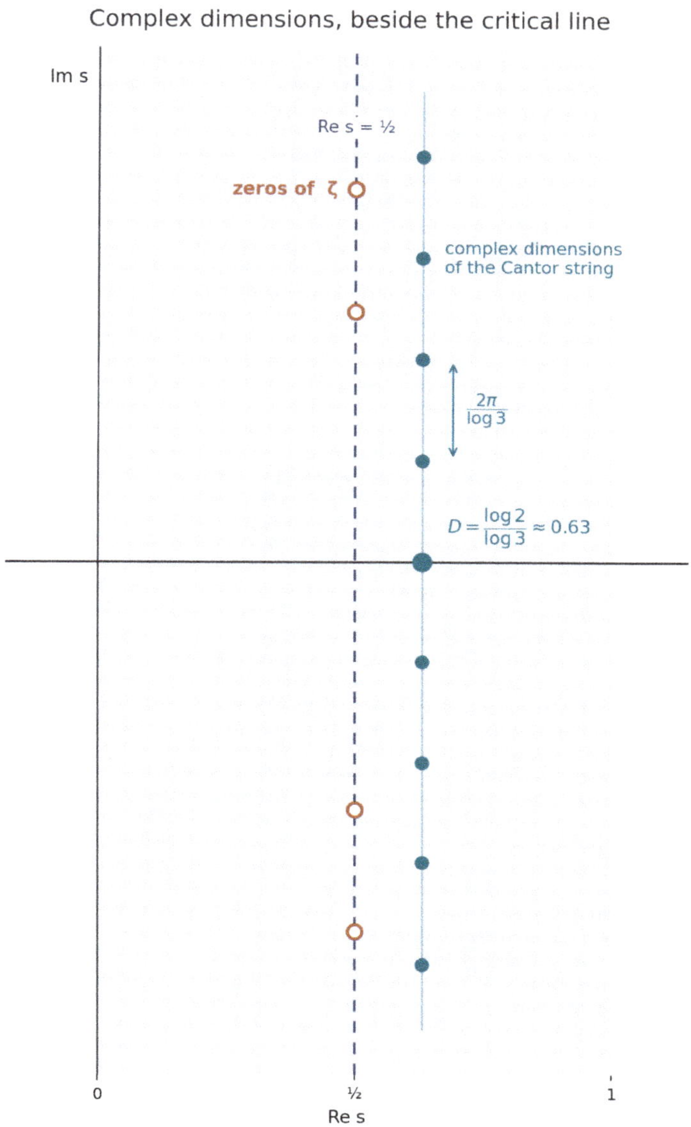

Figure A.1: Complex dimensions, beside the critical line. The Cantor string's complex dimensions—the poles of $1/(3^s - 2)$—form an exact vertical lattice on $\operatorname{Re} s = \log 2/\log 3$, its fractal dimension, spaced by $2\pi/\log 3$; they run an explicit formula for the string's geometry as the zeros of ζ run one for the primes. The lattice is perfectly periodic because the Cantor set is exactly self-similar—the zeros of ζ (on $\operatorname{Re} s = \frac{1}{2}$) are not a lattice, which is one way to say the primes are not. By Lapidus–Maier, $\frac{1}{2}$ is the lone dimension at which the inverse spectral problem can fail.

A.3 Two famous sharpenings that are false

The error-term family carries a warning the catalogue should never drop. Drop the ε from the Mertens bound and one gets the *Mertens conjecture*, $|M(x)| \leq \sqrt{x}$. It holds for every x anyone has ever computed, including the entire range of the verification here—and it is false, disproven by Odlyzko and te Riele in 1985 without exhibiting a single counterexample. Replace the Liouville bound by its tempting strong form and one gets *Pólya's conjecture*, $L(x) \leq 0$ for $x \geq 2$. It too holds throughout the verification's range, and it too is false: the first x with $L(x) > 0$ is 906,150,257. Both are the lure that dissolves, in its starkest form—a pattern exact across every accessible case and untrue past the horizon of computation. That the genuine equivalences keep the ε is not pedantry; it is the whole difference between a theorem and a coincidence.

A.4 Approaches, not equivalences: the spectral programme

Items $7, 9, 10, 13, 15$ are the heart of the companion volume's spectral chapter, and not one of them is a proven equivalence.

The Hilbert–Pólya conjecture (7) would *imply* RH—a self-adjoint operator has real eigenvalues—but the implication runs one way, and the reverse is vacuous in the worst sense: if RH holds, the diagonal operator on the zero heights is already such an operator, the made-to-order construction whose emptiness the spectral chapter dwells on. The random-matrix correspondence (9) is the reverse case: Montgomery's pair-correlation result *assumes* RH and describes the finer statistics of zeros taken to be on the line; the agreement with the GUE eigenvalue statistics, confirmed numerically by Odlyzko, is the strongest evidence in the subject and is evidence, not equivalence. Berry and Keating's $H = xp$ (10) is a conjectural model; Connes' noncommutative geometry (13) is a programme, whose genuine equivalence is the Weil positivity already listed; and the quantum-chaos correspondence (15) is an analogy. Each is serious. None is an iff.

A.5 An analogy, and a non-equivalence

The function-field mirror (14). The Riemann Hypothesis for varieties over finite fields is a theorem—proven by Weil in the curve case and by Deligne in general—and it is the deepest source of confidence in the classical Hypothesis. But it is an analogy in another category, a statement about a different class of zeta functions, and it neither implies nor is implied by RH for $\zeta(s)$. It belongs on the list as inspiration, not as equivalence.

The divisor bound (11). This one is simply miscatalogued. The Dirichlet divisor error $\Delta(x) = \sum_{n \leq x} d(n) - x \log x - (2\gamma - 1)x$ is known *unconditionally* to be $O(x^{1/3})$ (Voronoï), with the open conjecture $O(x^{1/4+\varepsilon})$ and a proven lower bound of order

$x^{1/4}$. The bound $O(x^{1/2+\varepsilon})$ written in the list is therefore weaker than what is already proven without any hypothesis—trivially true, equivalent to nothing. The verification shows $\Delta(x)$ staying far below $\sqrt{x}$ across the whole range. The divisor problem is a hard and separate question, independent of RH.

The Möbius sum (12). This entry needs splitting. At $\theta = 0$ the sum $\sum_{n\le x} \mu(n)e(n\theta)$ is exactly the Mertens function, and its $O(x^{1/2+\varepsilon})$ bound is item 3, a genuine equivalence. Uniform over rational θ it becomes a statement about μ in arithmetic progressions, equivalent to the *generalized* Riemann Hypothesis, not the classical one. And under its usual name, the "Möbius randomness principle" is Sarnak's disjointness conjecture—$\sum_{n\le x} \mu(n)f(n) = o(x)$ for f from any zero-entropy system—a far broader statement that is not equivalent to RH at all.

What this is and is not

What is genuine here is a substantial body of exact equivalences: the four error-term forms, Li's positivity, Nyman–Beurling and Báez-Duarte, the Laguerre–Pólya/de Bruijn–Newman characterization, Robin's and Lagarias's inequalities, the Riesz and Franel–Landau and Redheffer criteria, and Weil's positivity, and the Lapidus–Maier inverse spectral problem for fractal strings. Each is proven equivalent to the Hypothesis, and each is named here in its sharp form.

What it is not is a list of fifteen equivalences. The spectral programme would imply RH or already assumes it; the function-field Hypothesis is an analogy in another world; the divisor bound is trivially true; the Möbius entry is three different statements wearing one name. And two of the most natural-looking strengthenings, the Mertens and Pólya conjectures, are outright false while holding across every computable case—which is the appendix's sharpest lesson and the book's: a reformulation is only as good as its proof of equivalence, and a pattern that holds as far as the eye can see is, until proven, only a pattern. Sorting these by status is not bookkeeping. It is the whole discipline of the subject in miniature.

Verification

The script below, using only the Python standard library and no random seeds, confirms the genuine error-term equivalences in range, exhibits the two false sharpenings holding throughout that same range, checks Robin's inequality past its last exception, and shows the divisor error sitting far below the threshold that would make its bound meaningful.

```
#!/usr/bin/env python3
"""Verification for 'Reformulations and Equivalences of RH'.
Standard library only; deterministic (no seeds)."""
from math import log, isqrt
import array, time
t0=time.time()
N=10**6
```

```
GAMMA=0.5772156649015329
EG=2.718281828459045**GAMMA   # e^gamma

checks=0
def ok(cond,msg):
    global checks
    assert cond,"FAIL: "+msg
    checks+=1; print("PASS:",msg)

# ---- smallest-prime-factor sieve -> mu, Omega ----
spf=list(range(N+1))
i=2
while i*i<=N:
    if spf[i]==i:
        for j in range(i*i,N+1,i):
            if spf[j]==j: spf[j]=i
    i+=1
mu=array.array('b',[0])*(N+1); mu[1]=1
Om=array.array('b',[0])*(N+1)
for n in range(2,N+1):
    x=n; om=0; sq=True
    while x>1:
        p=spf[x]; c=0
        while x%p==0: x//=p; c+=1; om+=1
        if c>1: sq=False
    Om[n]=om%2
    mu[n]=0 if not sq else (1 if om%2==0 else -1)

# ---- divisor sieve -> sigma(n) and divisor count ----
sigma=array.array('q',[0])*(N+1)
dct=array.array('i',[0])*(N+1)
for d in range(1,N+1):
    for m in range(d,N+1,d):
        sigma[m]+=d; dct[m]+=1

# ===== genuine error-term equivalences hold in range (RH-consistent) =====
M=0; L=0; maxM=0; maxL=0; polya_ok=True; mert_ok=True
for x in range(1,N+1):
    M+=mu[x]; L+= 1 if Om[x]==0 else -1
    if abs(M)>maxM: maxM=abs(M)
    if abs(L)>maxL: maxL=abs(L)
    if abs(M) > isqrt(x): mert_ok=False        # Mertens conjecture |M(x)|<=sqrt(x)
    if x>=2 and L>0: polya_ok=False            # Polya conjecture L(x)<=0
ok(maxM < N**0.6 and maxL < N**0.6,
   f"M(x), L(x) stay O(x^0.6) to 10^6 (max|M|={maxM}, max|L|={maxL})  [RH-equivalent]")

# ===== two FALSE sharpenings, holding throughout this range =====
ok(mert_ok,
   "Mertens conj. |M(x)|<=sqrt(x) holds to 10^6 -- yet FALSE (Odlyzko-te Riele)")
ok(polya_ok,
   "Polya conj. L(x)<=0 holds to 10^6 -- yet FALSE (first sign change 906150257)")

# ===== Robin's inequality (equivalent to RH) holds past its last exception 5040 =====
robin_ok=True
for n in range(5041,N+1):
    if sigma[n] >= EG*n*log(log(n)): robin_ok=False; break
ok(robin_ok,
   "Robin sigma(n)<e^g n loglog n for 5040<n<=10^6 (equiv to RH; unproven beyond)")

# ===== the Dirichlet divisor problem is NOT RH-equivalent =====
# D(x)=sum d(n)= x log x + (2g-1)x + Delta(x); |Delta(x)| is O(x^{1/3}) UNCONDITIONALLY
# (Voronoi) -- far below x^{1/2+eps}, so the stated bound is trivially true.
D=0; maxDelta=0.0
```

```
for x in range(1,N+1):
    D+=dct[x]
    delta=D-(x*log(x)+(2*GAMMA-1)*x)
    if abs(delta)>maxDelta: maxDelta=abs(delta)
print(f"  divisor: max|Delta|={maxDelta:.0f}, sqrt(1e6)=1000, ratio {maxDelta/N**0.5:.3f}")
ok(maxDelta < N**0.5,
   "divisor error << sqrt(x) unconditionally: that bound is trivial, not RH-equivalent")

print(f"\nALL {checks} CHECKS PASSED   ({time.time()-t0:.1f}s)")
```

Exercises

The exercises are collected here, by chapter, rather than interrupting the text. Difficulty is marked [routine], [moderate], [hard], with [computational] for those that extend a chapter's verification program, [discussion] for those that ask for an argument in prose, and [conceptual] for those that ask for a precise statement of a concept or thesis. The recurring *two-part turn* referred to throughout is the pattern of every chapter: a method grown from the wheel climbs to the wall, and the lever that would cross it proves to be the very thing being sought.

Chapter 1 The Wheel and the Wall

1. [routine] Show that the number of residues coprime to 210 is $\prod_{p|210}(p-1) = 1 \cdot 2 \cdot 4 \cdot 6 = 48$, and that the density $48/210$ equals $\prod_{p|210}(1-1/p)$.

2. [routine] A *subprime* is a composite coprime to 210 whose least prime factor is at least 11. Find the smallest three.

3. [moderate] From the Euler product $\zeta(s) = \prod_p (1-p^{-s})^{-1}$, derive $1/\zeta(s) = \sum_n \mu(n)n^{-s}$, identifying μ as the multiplicative function this forces.

4. [moderate] Using the functional equation $\xi(s) = \xi(1-s)$ and $\overline{\xi(s)} = \xi(\bar{s})$, show the nontrivial zeros are symmetric about the line $\operatorname{Re} s = \frac{1}{2}$ and about the real axis, so they come in quadruples unless they sit on the line.

5. [discussion] The explicit formula writes $\psi(x)$ as x minus a sum over the zeros. Explain why this makes "the location of the zeros" and "the size of the error $\psi(x) - x$" two names for one thing, and why that circularity is the shape of every later chapter.

Chapter 2 One Floor Up: Ramanujan's τ

1. [routine] Verify Ramanujan's congruence $\tau(p) \equiv p^{11} + 1 \pmod{691}$ for $p = 2$, given $\tau(2) = -24$.

2. [moderate] Deligne's bound gives $|\tau(p)| \le 2p^{11/2}$. Writing $\tau(p)/p^{11/2} = 2\cos\theta_p$, compute θ_2 from $\tau(2) = -24$, and explain why $\theta_p \in [0, \pi]$ is the "angle on the circle" analogue of a zero on the critical line.

3. [moderate] Explain in what precise sense Deligne's theorem is "the Riemann Hypothesis proved one floor up": which object plays the rôle of ζ, which map plays the rôle of the zeros, and why $|\alpha| = p^{11/2}$ is the analogue of $\operatorname{Re} s = \frac{1}{2}$.

4. [discussion] The *local* Hypothesis for Δ is a theorem (Deligne); the *global* Hypothesis for $L(s,\Delta)$ is open. Distinguish the two statements and say why proving the first did not settle the second.

5. [hard] Sato–Tate (the θ_p are semicircle-distributed) is a theorem for Δ. Does it imply the global Hypothesis for $L(s,\Delta)$? Explain why a law for the *distribution* of the θ_p does not *locate* the zeros of $L(s,\Delta)$.

Chapter 3 The Oscillation Face

1. [computational] Extend the verification: with a Möbius sieve compute $M(x) = \sum_{n\le x}\mu(n)$ for $x = 10$ and $x = 100$, and check $|M(x)| \le \sqrt{x}$ in each case.

2. [moderate] Derive the Dirichlet series $\sum_n \lambda(n)n^{-s} = \zeta(2s)/\zeta(s)$ for the Liouville function $\lambda(n) = (-1)^{\Omega(n)}$.

3. [moderate] Each of $\pi-\mathrm{Li}$, $\psi-x$, M, and the Liouville sum is $O(x^{1/2+\varepsilon})$ if and only if RH. Using the explicit formula, explain why all four are "the same oscillation."

4. [discussion] The Mertens conjecture $|M(x)| \le \sqrt{x}$ is false, yet no counterexample is known below about 10^{16}. Reconcile "false" with "holds throughout all computation," and say what this teaches about numerical evidence for the Hypothesis.

5. [discussion] Pólya conjectured the Liouville sum stays ≤ 0 for $x \ge 2$; the first sign change is at 906,150,257. Why is such a distant counterexample consistent with the zeros picture, and why does "checked to 10^8" prove nothing?

Chapter 4 The Positivity Face

1. [routine] For a zero $\rho = \frac{1}{2} + it$ on the line, show $\big|1 - 1/\rho\big| = 1$ by computing $\big|1 - 1/(\frac{1}{2} + it)\big|^2$.

2. [moderate] Show the Möbius map $w = 1 - 1/s$ sends $\operatorname{Re} s = \frac{1}{2}$ to $|w| = 1$, $\operatorname{Re} s > \frac{1}{2}$ to $|w| < 1$, and $\operatorname{Re} s < \frac{1}{2}$ to $|w| > 1$.

3. [moderate] Li's coefficients are $\lambda_n = \sum_\rho\big[1 - (1 - 1/\rho)^n\big]$. Explain why on-line zeros ($|1 - 1/\rho| = 1$) keep the terms bounded and the $\lambda_n \ge 0$, while an off-line pair ($|1 - 1/\rho| > 1$) must eventually force some $\lambda_n < 0$.

4. [computational] Confirm $\lambda_1, \dots, \lambda_{10}$ are positive and increasing for the on-line zeros; then insert the off-line quadruple $\{0.1\pm i,\ 0.9\pm i\}$ and find the first n with $\lambda_n < 0$.

5. [discussion] Weil positivity ($W(g,g) \geq 0$ for all $g \Leftrightarrow$ RH) is the target of the Connes programme. Why is exhibiting positivity for many test functions not a proof, and how does this point at the operator wall of Chapter 8?

Chapter 5 The Completeness Face

1. [routine] Verify the bridge identity $\zeta(s) = \dfrac{s}{s-1} - s\displaystyle\int_1^\infty \{x\}\, x^{-s-1}\, dx$ at $s = 2$, where $\{x\}$ is the fractional part.

2. [moderate] Show $\langle 1, \rho_{1/k}\rangle = \frac{1}{k}\,(1 - \gamma + \ln k)$ for the building block $\rho_\theta(t) = \{\theta/t\}$ on $(0,1)$.

3. [moderate] Explain the Mellin-transform bridge: why the dilations $\{\theta/t\}$ carry $\zeta(s)/s$, so their closed span reaches "down to the rightmost zero," making completeness (reaching the constant 1) equivalent to RH.

4. [computational] Extend the chapter's computation: form the Nyman–Beurling distance d_N for $N = 2, \ldots, 8$, confirm it decreases, and compare its decay to $C/\log N$.

5. [discussion] Báez-Duarte showed integer dilations suffice (one sequence with $\|b_N\| \to 0 \Leftrightarrow$ RH). Why does the distance shrinking, even at a rate set by the zeros, not amount to a proof?

Chapter 6 The Extremal Face

1. [routine] Compute $\sigma(n)/n$ for the primorials $n = 2, 6, 30, 210$ and compare each with $e^\gamma \log\log n$; observe that 6 and 30 are among the twenty-seven Robin exceptions (ratio > 1), while 210 already satisfies the bound (ratio $0.918 < 1$).

2. [routine] Verify Lagarias's inequality $\sigma(n) \leq H_n + e^{H_n} \log H_n$ at $n = 1$ (with equality) and at $n = 2, 3, 4$ (strictly).

3. [moderate] Show $\sigma(n)/n = \prod_{p^a \| n}(1 + p^{-1} + \cdots + p^{-a})$ and explain why the maximizers use many small primes—the primorial cascades.

4. [computational] Extend the divisor sieve to find the largest $n \leq 10^6$ that violates Robin's inequality, and confirm there are exactly twenty-seven exceptions.

5. [discussion] The margin beneath Robin's ceiling is $\psi(y) - y$ in divisor costume. Explain why testing Robin on the champions is the same square-root cancellation as the oscillation face, so it offers no new route to the wall.

Chapter 7 The Regularity Face

1. [routine] For the toy $f_t(z) = z^2 + (c - 2t)$, find the zeros and show they are real iff $t \geq c/2$; identify the threshold $\Lambda = c/2$.

2. [routine] For $c = 0$ (so $\Lambda = 0$), show the zero is real at $t = 0$ and complex for every $t < 0$—the "only barely so" edge.

3. [computational] Confirm that $H_0(z) = \int \Phi(u)\cos(zu)\,du$ vanishes at $z = 2\gamma_1$ for the first Riemann ordinate (recall Φ is even, so evaluate at $|u|$).

4. [moderate] Rodgers and Tao proved $\Lambda \geq 0$, and RH $\Leftrightarrow \Lambda \leq 0$. Deduce RH $\Leftrightarrow \Lambda =$ 0 and explain Newman's remark that the Hypothesis, if true, is only barely so.

5. [discussion] The upper bounds on Λ descend toward 0 (Polymath15: 0.22). Explain why reaching 0 from above is not "almost proving RH" but *is* proving RH.

Chapter 8 The Neighbouring Walls

1. [computational] Unfold the first sixty zeros by $N(T)$, form the nearest-neighbour spacings, and show the mean is near 1 with few small gaps—the repulsion of GUE, not the clumping of Poisson.

2. [moderate] The semiclassical count for $H = xp$ reproduces $N(T) = \frac{T}{2\pi}\log\frac{T}{2\pi} - \frac{T}{2\pi} + \frac{7}{8}$. Evaluate $N(T)$ at $T = 163.03$ (the 60th ordinate) and compare with the count 60.

3. [routine] For $y^2 = x^3 + x + 1$ over $\mathbb{F}_{101}$, compute $a_p = p + 1 - \#E(\mathbb{F}_p)$ and check Hasse's bound $|a_p| \leq 2\sqrt{p}$; confirm $a_p^2 \leq 4p$, so the Frobenius eigenvalues lie on $|\alpha| = \sqrt{p}$.

4. [conceptual] The Lindelöf Hypothesis $\zeta(\frac{1}{2} + it) = O(t^{\varepsilon})$ is implied by RH and is weaker. Explain why subconvexity exponents creeping toward 0 do not reach it, and why even this weaker wall is open.

5. [discussion] The four neighbours form a ring: the statistics demand an operator, the operator would give RH and supply Weil positivity, and the function-field world owns the geometry that yields that positivity. Say what is missing, over $\mathbb{Z}$, on each side.

Chapter 9 The Twin Corridor

1. [computational] Compute the partial products $\prod_{3 \leq p \leq P}(1 - (p-1)^{-2})$ for growing P and watch them approach $C_2 = 0.6601618\ldots$; the twin constant is $2C_2$.

2. [moderate] Brun proved the sum of reciprocals of the twin primes converges, while $\sum_p 1/p$ diverges. What does the contrast say about how thinly the twins are spread?

3. [conceptual] State the parity barrier: a sieve cannot separate an even from an odd number of prime factors. Why does this stop the bounded-gap method one step short of gap two?

4. [routine] The bounded-gap record fell from 70,000,000 (Zhang) to 246 (Maynard–Tao), to 12 under Elliott–Halberstam, and to 6 under its generalized form. Explain why "bounded gaps" is weaker than "infinitely many twins."

5. [discussion] Chen's theorem—infinitely often p and $p+2$ are a prime and a number with at most two prime factors—sits at the edge of the sieve. Why is it the most the parity barrier allows, and why is gap two one parity step beyond?

Chapter 10 The Walls

1. [computational] Verify Lagarias's inequality for $1 \leq n \leq 100$ and identify the unique case of equality.

2. [computational] Build $N = pq$ from two large primes, then run trial division to 10^5 to confirm the sieve finds no factor though N was a single multiplication.

3. [conceptual] State the "one wall, three faces" thesis: for the Riemann Hypothesis, the twins, and factoring, name in each the local law that is known and the global fact that is walled.

4. [conceptual] Show that in each of the three walls the lever that would cross it equals the destination: controlling the zeros needs the zeros, breaking parity is isolating two primes, undoing multiplication is factoring.

5. [discussion] The recurring turn is "the structure you build is the thing you seek." Choose two faces from Part II and one neighbour from Part III and state the turn in each.

Solutions to the Exercises

These are answer sketches, not full write-ups: enough to grade against and to run a discussion from. Exercise numbers match the *Exercises* section at the back of the book. The recurring *two-part turn* is the pattern of every chapter—a method grown from the wheel climbs to the wall, and the lever that would cross it proves to be the very thing being sought.

Chapter 1 The Wheel and the Wall

1. By the Chinese remainder theorem the units mod 210 number $\prod_{p|210}(p-1) = 1 \cdot 2 \cdot 4 \cdot 6 = 48$. The density is $\frac{48}{210} = \frac{1}{2} \cdot \frac{2}{3} \cdot \frac{4}{5} \cdot \frac{6}{7} = \prod_{p|210}(1-1/p)$, each factor deleting the multiples of one prime.

2. $121 = 11^2$, $143 = 11 \cdot 13$, $169 = 13^2$. (Each is coprime to 210 with least prime factor ≥ 11; the next are $187 = 11 \cdot 17$ and $209 = 11 \cdot 19$.)

3. $1/\zeta(s) = \prod_p(1 - p^{-s})$. Expanding, n^{-s} appears with coefficient $(-1)^{\#\text{ distinct primes of } n}$ if n is squarefree and 0 otherwise—that is $\mu(n)$. Hence $1/\zeta(s) = \sum_n \mu(n)n^{-s}$.

4. $\xi(s) = \xi(1-s)$ sends a zero ρ to $1-\rho$ (reflection across $\mathrm{Re} = \frac{1}{2}$); real coefficients give $\overline{\xi(s)} = \xi(\bar{s})$, so $\bar{\rho}$ is also a zero. Thus zeros occur as $\{\rho, 1-\rho, \bar{\rho}, 1-\bar{\rho}\}$—a genuine quadruple off the line, collapsing to the conjugate pair $\{\rho, \bar{\rho}\}$ when $\rho = 1 - \bar{\rho}$, i.e. on the line.

5. Each zero contributes a term of size $x^{\mathrm{Re}\,\rho}$ to $\psi(x) - x = -\sum_\rho x^\rho/\rho - \cdots$, so the *size* of the error and the *real part* of the zeros are one datum. To bound the error one must place the zeros, and the error is the only handle on them: the lever is the destination. Every face repeats this.

Chapter 2 One Floor Up: Ramanujan's τ

1. $\tau(2) = -24$; $2^{11} + 1 = 2049 \equiv 2049 - 2 \cdot 691 = 667 \pmod{691}$; and $-24 \equiv 691 - 24 = 667$. Equal.

2. $\tau(2)/2^{11/2} = -24/45.255 = -0.5304 = 2\cos\theta_2$, so $\cos\theta_2 = -0.2652$ and $\theta_2 \approx 1.84$ rad ($\approx 105°$). The conjugate eigenvalues are $p^{11/2}e^{\pm i\theta_p}$ on the circle of

radius $p^{11/2}$; θ_p is the angular position, the GL(2) analogue of a zero's height on the line.

3. The object is $L(s, \Delta)$, with local Euler factor $(1 - \tau(p)p^{-s} + p^{11-2s})^{-1}$ whose inverse roots $\alpha, \beta = p^{11/2}e^{\pm i\theta_p}$ are the Frobenius eigenvalues. The local Hypothesis is $|\alpha| = p^{11/2}$, the fixed-modulus condition that mirrors a fixed real part. Deligne proved it as eigenvalues of Frobenius on the cohomology of the associated (Kuga–Sato) variety—the Weil conjectures.

4. Local: each Euler factor's roots have $|\alpha| = p^{11/2}$ (a bound on the coefficients, proven). Global: all nontrivial zeros of the continued $L(s, \Delta)$ lie on its critical line (open). The first is a statement in the region of convergence; the second is about the analytic continuation. Bounding coefficients does not locate continued zeros.

5. No. Sato–Tate equidistributes the angles θ_p by the semicircle law—a statement about their *average distribution.* The zeros of $L(s, \Delta)$ are a continuation feature; equidistribution constrains average size, not the real part of any zero. Distribution is not location.

Pitfall: *the recurring error is to read Deligne's* $|\tau(p)| \leq 2p^{11/2}$ *(local, proved) as the global Hypothesis for* $L(s, \Delta)$. *Ex. 4 exists to separate them.*

Chapter 3 The Oscillation Face

1. μ on $1, \ldots, 10$ is $1, -1, -1, 0, -1, 1, -1, 0, 0, 1$, summing to $M(10) = -1$, and $|-1| \leq \sqrt{10} \approx 3.16$. Likewise $M(100) = 1 \leq 10$.

2. λ is completely multiplicative with $\lambda(p) = -1$, so the local factor is $\sum_{k\geq 0}(-1)^k p^{-ks} = (1 + p^{-s})^{-1}$. Hence $\sum_n \lambda(n)n^{-s} = \prod_p(1 + p^{-s})^{-1} = \prod_p \frac{1-p^{-s}}{1-p^{-2s}} = \zeta(2s)/\zeta(s)$.

3. The explicit formula writes each of $\pi - \mathrm{Li}$, $\psi - x$, M, and the Liouville sum as a sum over the zeros, term size $x^{\mathrm{Re}\,\rho}$. So each is $O(x^{1/2+\varepsilon})$ iff every $\mathrm{Re}\,\rho = \frac{1}{2}$ (RH), and "the oscillation" is literally the same zero-sum, dressed by μ, λ, or Λ.

4. The disproof (Odlyzko–te Riele) is non-constructive and the first counterexample is astronomically large, so $|M| \leq \sqrt{x}$ can survive every feasible computation while being false. Numerical agreement to any reachable height is not proof; the zeros' fine conspiracies surface only far out.

5. The Liouville sign is governed by the same zero-sum; a slow bias can persist to $\sim 9 \times 10^8$ before flipping. "Checked to 10^8" certifies only the range tested, and the relevant scales for such questions can be vastly larger (compare Skewes's number for $\pi - \mathrm{Li}$).

Chapter 4 The Positivity Face

1. $|1-1/\rho|^2 = |\rho-1|^2/|\rho|^2$. For $\rho = \frac{1}{2}+it$, $|\rho-1|^2 = \frac{1}{4}+t^2 = |\rho|^2$, so the ratio is 1.

2. $|w|^2 = |s-1|^2/|s|^2 = \dfrac{(\sigma-1)^2+t^2}{\sigma^2+t^2}$ for $s=\sigma+it$. Then $|w|<1 \Leftrightarrow (\sigma-1)^2 < \sigma^2 \Leftrightarrow \sigma > \frac{1}{2}$, with equality at $\sigma = \frac{1}{2}$ and $|w|>1$ for $\sigma < \frac{1}{2}$.

3. Write $w_\rho = 1 - 1/\rho$. On the line $|w_\rho| = 1$, so w_ρ^n stays on the unit circle and $1-w_\rho^n$ is bounded; summed over conjugate pairs the contributions are ≥ 0 and $\lambda_n \geq 0$. An off-line zero with Re $< \frac{1}{2}$ has $|w_\rho| > 1$, so w_ρ^n grows geometrically and $-w_\rho^n$ eventually drives some $\lambda_n < 0$.

4. The on-line $\lambda_1, \ldots, \lambda_{10}$ come out positive and increasing $(0.019, 0.076, \ldots, 1.87)$. Inserting $\{0.1 \pm i,\, 0.9 \pm i\}$ (with $|1-1/(0.1+i)| = 1.339 > 1$) turns λ_n negative first at $n=7$.

5. $W(g,g) \geq 0$ for *all* g is RH; any finite or explicit family of g tests only those directions. The universal statement is the spectral side of a trace formula—the missing operator of Chapter 8. Positivity on samples is evidence, not the theorem.

Pitfall: *in Ex. 1 students expand numerically; the clean route is* $|1-1/\rho|^2 = |\rho-1|^2/|\rho|^2$, *after which the line* $\rho = \frac{1}{2}+it$ *gives* $|\rho-1| = |\rho|$ *in one step.*

Chapter 5 The Completeness Face

1. At $s=2$, $\dfrac{s}{s-1} = 2$, and the identity reads $\zeta(2) = 2 - 2\int_1^\infty \{x\}x^{-3}dx$. Computing the integral per unit interval gives ≈ 0.1776, so the right side is $2 - 2(0.1776) = 1.6449 = \pi^2/6$. Both sides agree.

2. With $u = \theta/t$ $(\theta = 1/k)$, $\langle 1, \rho_{1/k}\rangle = \int_0^1 \{\theta/t\}dt = \theta\int_\theta^\infty \{u\}u^{-2}du$. Split at 1: on $[\theta,1]$, $\{u\} = u$ gives $\int_\theta^1 du/u = \ln k$; and $\int_1^\infty \{u\}u^{-2}du = 1-\gamma$. Hence $\langle 1, \rho_{1/k}\rangle = \frac{1}{k}(1-\gamma+\ln k)$.

3. The dilations have Mellin transform proportional to $\zeta(s)/s$, so a function lies in their closed span only where that transform allows—reaching every point to the right of the rightmost zero. Reaching the constant 1 therefore requires no zero with Re $> \frac{1}{2}$, i.e. RH.

4. The distances decrease $0.181 \to 0.083 \to 0.055 \to 0.049 \to 0.048 \to \cdots$; a fit against $C/\log N$ matches, with C governed by $\sum 1/|\rho|^2$.

5. Báez-Duarte's $\|b_N\| \to 0$ is equivalent to RH, and the very rate of approach is set by the zeros. Observing $d_N \to 0$ numerically is not proving the limit is 0; the limit *being* 0 is the Hypothesis.

Chapter 6 The Extremal Face

1. $\sigma/n = 1.5$, 2, 2.4, 2.743 for $n = 2, 6, 30, 210$. At $n = 210$, $e^{\gamma} \log\log n = 1.781 \cdot 1.677 = 2.987$, so the Robin ratio is $2.743/2.987 = 0.918 < 1$.

2. $n = 1$: $\sigma = 1$, $H_1 = 1$, RHS $= 1 + e \cdot \ln 1 = 1$—equality. $n = 2, 3, 4$: RHS $\approx 3.32, 5.62, 7.98$ exceed $\sigma = 3, 4, 7$.

3. $\sigma(n)/n = \prod_{p^a \| n}(1 + p^{-1} + \cdots + p^{-a}) < \prod_{p|n} \frac{p}{p-1}$. Each prime contributes near $p/(p-1)$, largest for the smallest p; so maximizers take the smallest primes with descending exponents—the primorial cascades (super/colossally abundant numbers).

4. Extending the divisor sieve to 10^6: the last violator is $n = 5040$ and the exceptions number exactly twenty-seven.

5. For a colossally abundant n, $\sigma(n)/n \approx \prod_{p \le y}(1 - 1/p)^{-1} \sim e^{\gamma} \log y$ (Mertens, exact), with $\log y \sim \log\log n$; the residual margin is $\sum_{p \le y} \log p - y = \psi(y) - y$, the Chebyshev error. So Robin's truth past 5040 is the same $\sqrt{y}$ cancellation as the oscillation face—one wall in divisor costume, no new route.

Pitfall: *in Ex. 1 the threshold $n > 5040$ is not a consequence of $\log\log n$ being small for small primorials. The primorials 6 and 30 genuinely violate Robin's inequality (ratio > 1), as do the rest of the twenty-seven exceptional values up to 5040; the bound holds for every $n > 5040$ because 5040 is the largest exception, not because of any $\log\log n$ size effect. Among the primorials, 210 is the first to satisfy the bound (ratio $0.918 < 1$).*

Chapter 7 The Regularity Face

1. $f_t(z) = z^2 + (c - 2t)$ has zeros $z = \pm\sqrt{2t - c}$, real iff $2t - c \ge 0$, i.e. $t \ge c/2$. So $\Lambda = c/2$.

2. $c = 0$: zeros $\pm\sqrt{2t}$. At $t = 0$ a real double zero; for $t < 0$, $\pm i\sqrt{2|t|}$, complex. Real exactly at and after $t = 0$, complex for every $t < 0$.

3. $H_0(z) = \int \Phi(u) \cos(zu)\, du$ with Φ even (evaluate at $|u|$, $n = 1 \ldots 6$); its first sign change falls at $z = 2\gamma_1 = 28.27$, matching the first ordinate.

4. $\Lambda \ge 0$ (Rodgers–Tao) and RH $\Leftrightarrow \Lambda \le 0$ force RH $\Leftrightarrow \Lambda = 0$. Since $\Lambda \ge 0$ always, RH puts Λ exactly at its floor: cool the flow by any amount and a zero leaves the line—"only barely so."

5. Reaching 0 from above means proving the flow is already real-rooted at $t = 0$, which *is* the Hypothesis. Each bound $\frac{1}{2} \to \cdots \to 0.22$ narrows the bracket, but the endpoint 0 is RH itself; there is no "almost."

Chapter 8 The Neighbouring Walls

1. Unfolding by $N(T)$ gives mean spacing ≈ 0.999 and only $\approx 3.4\%$ of gaps below half the mean, against 39% for Poisson: repulsion, the GUE signature.

2. $N(163.03) = \frac{163.03}{2\pi}\log\frac{163.03}{2\pi} - \frac{163.03}{2\pi} + \frac{7}{8} = 25.948\cdot 3.256 - 25.948 + 0.875 \approx 59.4$, against the count 60—agreeing within the small $S(T)$ fluctuation.

3. For $y^2 = x^3 + x + 1$ over $\mathbb{F}_{101}$ the point count is 105, so $a_p = 101 + 1 - 105 = -3$. Then $|a_p| = 3 \le 2\sqrt{101} \approx 20.1$ and $a_p^2 = 9 \le 4\cdot 101 = 404$, so the Frobenius eigenvalues lie on $|\alpha| = \sqrt{101}$.

4. Each subconvexity gain is hard-won and bounded away from 0; t^ε is the infimum, attained by no finite exponent. And Lindelöf, though implied by RH, is equivalent to an open statement about the horizontal spread of the zeros, so it stands unproved on its own.

5. Missing over $\mathbb{Z}$: the operator (none built; xp unregularized), the geometry ($\operatorname{Spec}\mathbb{Z}$ is no curve over a field—no surface, no Frobenius, no $\mathbb{F}_1$), and hence the positivity (the Weil functional is not shown ≥ 0). The statistics are evidence only, and Lindelöf is weaker yet open. Each side lacks the one object that would close it.

Chapter 9 The Twin Corridor

1. The partial products $\prod_{3\le p\le P}(1-(p-1)^{-2})$ decrease to $C_2 = 0.6601618\ldots$, agreeing to several digits well before P reaches the thousands; the twin constant is $2C_2 = 1.3203\ldots$.

2. $\sum 1/p$ diverges while $\sum 1/(\text{twin primes})$ converges (to Brun's constant ≈ 1.902). The twins are therefore far thinner than the primes—almost all primes are not twins.

3. A sieve's weights cannot read $\Omega(n) \bmod 2$, so it cannot certify "exactly one prime factor" against "two." The bounded-gap method places a prime in a short window but cannot force the second member to be prime rather than a product; gap two needs precisely that parity decision.

4. Bounded gaps give infinitely many pairs within a fixed distance H (now 246, 12 under Elliott–Halberstam, and 6 under its generalized form), not distance exactly 2, and not even a fixed recurring gap. Infinitely many twins is the pinned case $H = 2$, which the sieve cannot reach.

5. Chen's theorem extracts everything up to parity—p with $p+2$ prime or a product of two primes—but cannot delete the second alternative, since that deletion is a parity determination. Gap two is exactly that forbidden step.

Pitfall: *students sometimes claim Chen "almost proves" the twins; stress that the remaining step is not small but is the parity barrier itself.*

Chapter 10 The Walls

1. Lagarias's inequality holds for every $1 \le n \le 100$; equality occurs only at $n = 1$.

2. With $N = pq$ a product of two large primes, trial division to 10^5 returns no factor, though N was formed by one multiplication—easy to make, hard to unmake.

3. **RH:** local law = the wheel and Euler product (residues, densities), known; global fact = the real parts of the zeros, walled. **Twins:** local law = the singular series $2C_2$ (predicted density), known; global = whether the count is infinite, walled. **Factoring:** local law = multiplication and primality (AKS), easy; global = recovering the factors, walled.

4. Controlling the zeros' real parts needs the prime error, which *is* the zeros (explicit formula); isolating two primes is removing the semiprime alternative, which *is* a parity determination; factoring cheaply is undoing a multiplication, which *is* factoring. In each, lever = destination.

5. Open discussion. For instance: positivity (a positive λ_n sequence $\Leftrightarrow$ no off-line zero, but checking needs the zeros), completeness (the span reaches 1 $\Leftrightarrow$ no zero right of the line, at a rate the zeros set), and the operator neighbour (an operator with the zeros as spectrum would give RH, but building it independently is the wall). Each: the structure built is the thing sought.

Glossary

AKS primality test A deterministic polynomial-time algorithm (Agrawal–Kayal–Saxena, 2004) that decides whether a number is prime; it shows that recognising primes is computationally easy.

a_p For an elliptic curve over $\mathbb{F}_p$, the quantity $a_p = p + 1 - \#E(\mathbb{F}_p)$. Hasse's bound $|a_p| \leq 2\sqrt{p}$ is the Riemann Hypothesis for the curve.

Báez-Duarte criterion A strengthening (2003) of Nyman–Beurling using only integer dilations: RH holds iff one explicit sequence of L^2 distances tends to zero.

Berry–Keating A proposed dynamics—the quantisation of the classical Hamiltonian $H = xp$—whose semiclassical level count reproduces $N(T)$; a candidate for the Hilbert–Pólya operator.

Brun's theorem The sum of the reciprocals of the twin primes converges (Brun, 1919), so the twins are sparse; Brun's sieve is the technique behind it.

Chebyshev functions $\theta(x) = \sum_{p \leq x} \log p$ and $\psi(x) = \sum_{p^k \leq x} \log p$. The error $\psi(x) - x$ is the prime-counting fluctuation governed by the zeros.

Chen's theorem Infinitely often p and $p+2$ are a prime and a number with at most two prime factors (Chen, 1973): the sieve pressed to its very edge.

Colossally abundant number An integer maximising $\sigma(n)/n^{1+\varepsilon}$; together with the superabundant numbers these are the extremal candidates for Robin's inequality, built from small primes.

Critical line / strip The line $\operatorname{Re} s = \frac{1}{2}$ and the strip $0 < \operatorname{Re} s < 1$ in which the nontrivial zeros of ζ lie; RH asserts they all lie on the line.

de Bruijn–Newman constant Λ The threshold time in the heat flow on Ξ: the flow is real-rooted iff $t \geq \Lambda$. RH $\Leftrightarrow \Lambda \leq 0$; Rodgers–Tao proved $\Lambda \geq 0$, so RH $\Leftrightarrow \Lambda = 0$. (Distinct from von Mangoldt's $\Lambda(n)$.)

Deligne's bound The proof (1974) of the Ramanujan bound $|\tau(p)| \leq 2p^{11/2}$—the Riemann Hypothesis for the associated variety—the local Hypothesis "one floor up."

Dirichlet character χ A periodic, completely multiplicative function modulo q; it builds the L-function $L(s, \chi)$.

Dirichlet divisor problem Bounding $\Delta(x) = \sum_{n \le x} d(n) - x(\log x + 2\gamma - 1)$, conjectured $O(x^{1/4+\varepsilon})$; a size-and-exponent cousin of the Lindelöf Hypothesis.

Euler–Mascheroni constant γ The limit of $H_n - \log n \approx 0.5772$, appearing in Robin's inequality and in Mertens' theorem.

Euler product $\zeta(s) = \prod_p (1 - p^{-s})^{-1}$: the wheel made into a function, binding the primes to ζ.

Explicit formula Von Mangoldt's identity writing $\psi(x)$—the primes—as x minus a sum over the zeros ρ; the hinge that carries information from the zeros to the primes.

Frobenius The map whose eigenvalues α determine a curve's zeta function over $\mathbb{F}_q$; the condition $|\alpha| = q^{1/2}$ is the function-field Riemann Hypothesis.

Functional equation The symmetry $\xi(s) = \xi(1-s)$ of the completed zeta function, reflecting the critical strip about $\operatorname{Re} s = \frac{1}{2}$.

GUE The Gaussian Unitary Ensemble of random Hermitian matrices, whose eigenvalue spacings (pair correlation $1 - (\sin \pi x / \pi x)^2$) match those of the Riemann zeros.

Harmonic number H_n $H_n = \sum_{k \le n} 1/k = \log n + \gamma + o(1)$; the elementary ingredient of Lagarias's criterion.

Hasse bound $|a_p| \le 2\sqrt{p}$ for elliptic curves over $\mathbb{F}_p$ (Hasse, 1930s): the Riemann Hypothesis for an elliptic curve, and a theorem.

Hilbert–Pólya conjecture The hope that the ordinates of the zeros are the eigenvalues of a self-adjoint operator, which would force them real and prove RH.

***L*-function** $L(s, \chi)$ The series $\sum \chi(n) n^{-s} = \prod (1 - \chi(p) p^{-s})^{-1}$; the family the wheel generates. Non-vanishing $L(1, \chi) \ne 0$ gives primes in arithmetic progressions.

Lagarias's criterion $\sigma(n) \le H_n + e^{H_n} \log H_n$ for all $n \ge 1 \Leftrightarrow$ RH (Lagarias, 2002): the most elementary statement of the Hypothesis.

Laguerre–Pólya class The real entire functions that are limits of real-rooted real polynomials; RH $\Leftrightarrow \Xi$ belongs to this class.

Level repulsion The vanishing probability of nearly-coincident zeros, the signature of GUE statistics and the opposite of independent (Poisson) points.

Li's criterion RH $\Leftrightarrow \lambda_n = \sum_\rho [1 - (1 - 1/\rho)^n] \ge 0$ for all $n \ge 1$ (Li, 1997); the Möbius map $w = 1 - 1/\rho$ sends the critical line exactly to the unit circle.

Li coefficients λ_n The numbers appearing in Li's criterion (distinct from Liouville's $\lambda(n)$).

Lindelöf Hypothesis $\zeta(\frac{1}{2}+it) = O(t^{\varepsilon})$; implied by RH, strictly weaker, and still open.

Liouville function $\lambda(n)$ $(-1)^{\Omega(n)}$, the parity of the number of prime factors; its summatory function $L(x)$ underlies the (false) Pólya conjecture. A sieve cannot see its sign.

Mellin transform The transform carrying the dilations $\{\theta/t\}$ to $\zeta(s)/s$ in the Nyman–Beurling criterion.

Mertens function $M(x)$ $\sum_{n\le x}\mu(n)$; the bound $M(x) = O(x^{1/2+\varepsilon})$ is equivalent to RH. The stronger Mertens conjecture $|M| \le \sqrt{x}$ is false (Odlyzko–te Riele, 1985).

Möbius function $\mu(n)$ Zero if n has a squared factor, else $(-1)^{\#\text{primes}}$; it gives $1/\zeta(s) = \sum \mu(n)n^{-s}$.

$N(T)$ The Riemann–von Mangoldt count of zeros up to height T: $\frac{T}{2\pi}\log\frac{T}{2\pi} - \frac{T}{2\pi} + \frac{7}{8} + \cdots$; the spectral staircase.

Number field sieve The fastest known classical factoring algorithm, running in time sub-exponential in the number of digits.

Nyman–Beurling criterion RH $\Leftrightarrow$ the constant 1 lies in the closed span of the dilations $\{\theta/t\}$ in $L^2(0,1)$: a completeness statement.

Parity barrier Selberg's obstruction: a sieve cannot distinguish an even from an odd number of prime factors, which stops the bounded-gap method one step short of gap two.

Primorial A product of the first primes $(2, 6, 30, 210, \dots)$; the wheel's modulus 210 is $2\cdot3\cdot5\cdot7$, and the extremal numbers of Robin's face are primorial cascades.

Riemann Hypothesis (RH) The conjecture that every nontrivial zero of ζ has real part exactly $\frac{1}{2}$.

Riemann zeta function $\zeta(s)$ $\sum n^{-s}$ for $\operatorname{Re} s > 1$, continued to the plane; its non-trivial zeros encode the distribution of the primes.

Robin's inequality $\sigma(n) < e^{\gamma} n \log\log n$ for all $n > 5040 \Leftrightarrow$ RH (Robin, 1984); below the threshold the inequality fails at exactly twenty-seven integers, the largest being 5040.

Sato–Tate The semicircle law for the Frobenius angles (and the angles of τ); proven for elliptic curves and for Δ.

Shor's algorithm A quantum algorithm (1994) that factors in polynomial time, so the factoring wall is not absolute—though it stands classically.

Singular series The local product (such as $2C_2$ for the twins) predicting the density of a prime pattern; it is drawn directly from the wheel.

Subconvexity Bounds on $|\zeta(\frac{1}{2}+it)|$ below the convexity exponent $\frac{1}{4}$ (Weyl $\frac{1}{6}$, and on toward $\frac{13}{84}$), creeping toward the Lindelöf value.

Subprime In this book, a composite coprime to 210 whose least prime factor is at least eleven —a wheel survivor that is not prime; the smallest is $121 = 11^2$.

Sum of divisors $\sigma(n)$ $\sum_{d|n} d$; the function of the extremal face (Robin and Lagarias).

Twin constant $2C_2$ $2\prod_{p>2}\left(1-\frac{1}{(p-1)^2}\right) = 1.3203\ldots$, the singular series for the twin primes.

von Mangoldt function $\Lambda(n)$ $\log p$ if $n = p^k$, else 0; it gives $\psi(x) = \sum_{n\le x}\Lambda(n)$. (Distinct from the de Bruijn–Newman constant Λ.)

Weil conjectures The Riemann Hypothesis for varieties over finite fields; proven by Weil (curves, 1948) and Deligne (the general case, 1974).

Weil positivity The positivity of an explicit functional, equivalent to RH and the target of the Connes programme.

Wheel This book's name for the structure of residues modulo $210 = 2\cdot 3\cdot 5\cdot 7$: the forty-eight survivors coprime to it, which encode the primes' local law.

$\xi(s)$ **and** $\Xi(z)$ The completed, entire zeta function with $\xi(s) = \xi(1-s)$, and its symmetric form $\Xi(z) = \xi(\frac{1}{2}+iz)$, the Fourier transform of the heat kernel Φ.

Bibliography

This is a guide to the sources behind the book, grouped by the part of the story each belongs to. It is a reading list, not a complete record; the aim is to point the reader to the original results named in the text.

The zeta function and the prime number theorem

- B. Riemann, *On the number of primes less than a given magnitude* (1859). The founding memoir: $\zeta(s)$, the functional equation, the explicit formula, and the Hypothesis.
- J. Hadamard (1896) and C.-J. de la Vallée Poussin (1896), independent proofs of the prime number theorem from the absence of zeros on $\operatorname{Re} s = 1$.
- P. G. L. Dirichlet (1837), primes in arithmetic progressions; the L-functions and $L(1, \chi) \neq 0$.
- H. von Mangoldt (1895), the explicit formula for $\psi(x)$.
- L. Schoenfeld (1976), explicit bounds for $\pi(x)$ and $\psi(x)$ under the Hypothesis.

One floor up: Ramanujan's τ

- S. Ramanujan (1916), the function $\tau(n)$ and its conjectured properties.
- L. J. Mordell (1917), the multiplicativity of τ.
- P. Deligne (1974), the Ramanujan bound $|\tau(p)| \leq 2p^{11/2}$, via the Weil conjectures. numdam.org
- L. Clozel, M. Harris, N. Shepherd-Barron, R. Taylor and collaborators (2008–2011), the Sato–Tate theorem.

The five faces (equivalences of the Hypothesis)

- X.-J. Li (1997), the positivity criterion $\lambda_n \geq 0$ (doi:10.1006/jnth.1997.2137); E. Bombieri and J. Lagarias (1999), complements and the arithmetic form of the λ_n. doi:10.1006/jnth.1999.2392

- A. Weil (1952), the explicit formula and the positivity equivalent to the Hypothesis.
- B. Nyman (1950) and A. Beurling (1955), the completeness criterion in $L^2(0,1)$; L. Báez-Duarte (2003), the integer-dilation strengthening.
- G. Robin (1984), the inequality $\sigma(n) < e^{\gamma} n \log\log n$; T. H. Gronwall (1913), the maximal order of $\sigma(n)$; J. Lagarias (2002), the elementary harmonic-number equivalent.
- N. G. de Bruijn (1950) and C. M. Newman (1976), the heat flow and the constant Λ; G. Csordas, W. Norfolk and R. Varga, the kernel and its flow; B. Rodgers and T. Tao (2020), $\Lambda \geq 0$; the Polymath15 project (2019), $\Lambda \leq 0.22$.

The neighbouring walls

- H. L. Montgomery (1973), the pair correlation of the zeros; A. M. Odlyzko (1987), the numerical zeros and their GUE statistics; J. Keating and N. Snaith (2000), random-matrix moments of ζ.
- D. Hilbert and G. Pólya (folklore, c. 1910s), the spectral conjecture; M. Berry and J. Keating (1999), $H = xp$ (doi:10.1137/S0036144598347497); A. Connes (1999), the trace formula on the adèle class space (doi:10.1007/s000290050042).
- G. H. Hardy (1914), infinitely many zeros on the line; A. Selberg (1942), N. Levinson (1974), and J. B. Conrey (1989), positive proportions of zeros on the line.
- E. Lindelöf, the size hypothesis; J. Bourgain and others, subconvexity bounds.
- A. Weil (1948), the Riemann Hypothesis for curves over finite fields; H. Hasse (1936), the elliptic-curve bound; P. Deligne (1974, 1980), the Weil conjectures in general.

The twin corridor and the sieve

- G. H. Hardy and J. E. Littlewood (1923), the prime k-tuple conjecture and the singular series. doi:10.1007/BF02403921
- V. Brun (1919), the convergence of the twin reciprocals and the sieve that proves it.
- J.-R. Chen (1973), the almost-twin theorem; A. Selberg, the parity problem.
- D. Goldston, J. Pintz and C. Yıldırım (2009), small gaps between primes; Y. Zhang (2013), bounded gaps (doi:10.4007/annals.2014.179.3.7); J. Maynard (2015) and the Polymath8 project, the gap 246 (doi:10.4007/annals.2015.181.1.7).

Factoring and computation

- M. Agrawal, N. Kayal and N. Saxena (2004), *PRIMES is in P.* doi:10.4007/annals.2004.160.781
- P. Shor (1994/1997), polynomial-time quantum factoring. doi:10.1109/SFCS.1994.365700
- A. K. Lenstra and H. W. Lenstra, Jr. (eds., 1993), the number field sieve.

Cautions: conjectures that failed

- A. M. Odlyzko and H. J. J. te Riele (1985), the disproof of the Mertens conjecture. doi:10.1515/crll.1985.357.138
- C. B. Haselgrove (1958), the disproof of the Pólya conjecture.

www.ingramcontent.com/pod-product-compliance
Lightning Source LLC
LaVergne TN
LVHW070855210926
848482LV00016B/942
* 9 7 9 8 9 9 6 7 3 2 5 6 2 *